Computational analysis of layered regulatory mechanisms in embryonic stem cells

Johannes Meisig

The illustration on the cover is composed of Kepler's description of planetary orbits by layers of Platonic solids in his work Mysterium Cosmographicum and the X:A ratio of ancestral genes in male stem cells (Fig 2.8).

Bibliografische Information der Deutschen Nationalbibliothek

Die Deutsche Nationalbibliothek verzeichnet diese Publikation in der Deutschen Nationalbibliografie; detaillierte bibliografische Daten sind im Internet über http://dnb.d-nb.de abrufbar.

Zugl.: Diss.: Humboldt-Universität zu Berlin, 2015

ISBN 978-3-8325-4242-9

Logos Verlag Berlin GmbH
Comeniushof, Gubener Str. 47,
10243 Berlin
Tel.: +49 (0)30 42 85 10 90
Fax: +49 (0)30 42 85 10 92
INTERNET: http://www.logos-verlag.de

Contents

1 Introduction

Cells in the early mammalian embryo have the unique ability to give rise to all somatic and germ line tissues. *In vivo*, this naive state, called pluripotency, is only transient. Pluripotent cells differentiate and give rise to individual tissues as development proceeds. In contrast to this, the pluripotent state can be captured and propagated *in vitro* by culturing embryonic stem cells in a chemically defined medium (Ying et al., 2008). Using differentiation cues, embryonic stem cells can be forced to exit the pluripotent state and differentiate into certain lineages (Kalkan and Smith, 2014). Thus, employing embryonic stem cells, we can study the maintenance of and the exit from the naive state in the laboratory.

A major focus of studies of this process have been the essential genes that have to be present to maintain the naive state and the signals that stabilize or destabilize this state. These essential genes form the core of a network that regulates maintenance of and departure from the naive state (Dunn et al., 2014). Characterizing this pluripotency-associated network and its dynamics is one of the key obstacles for understanding differentiation of embryonic stem cells. Advances in this direction will help to explain how the crucial decision to irreversibly exit from the pluripotent state is regulated.

Once cells are committed to departing from pluripotency, wide-ranging changes in the transcriptome are triggered by multiple regulatory mechanisms (Buecker et al., 2014; Kalkan and Smith, 2014). These mechanisms are interacting with each other and with the pluripotency-associated network. Among the mechanisms shaping the transcriptome are gene regulation by transcription factors, epigenetic mechanisms such as DNA methylation and histone modifications as well as signaling pathways (Hackett and Surani, 2014; Ng and Surani, 2011). In this work, I will focus on three regulatory mechanisms that shape gene expression during differentiation: dosage balancing of the X-chromosome between the sexes, spatial localization of chromatin in the nucleus and gene regulation by transcription factors.

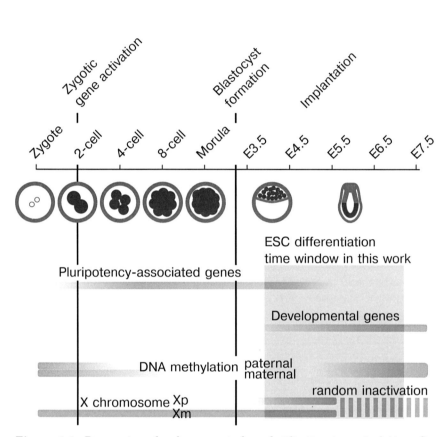

Figure 1.1: Progression of embryogenesis from fertilization to gastrulation. Selected processes taking place during this time are indicated by a white to yellow gradient, where white means minimal concentration/activity and yellow maximal concentration/activity. During blastocyst formation, the inner cell mass forms around embryonic day 3.5 (E3.5). Following E3.5, the processes indicated are only shown for the inner cell mass, after E5.5 only for the epiblast (both indicated in red). Embryonic stem cells that are harvested from the inner cell mass can be cultured and differentiated *in vitro*. In this work, the transcriptome of embryonic stem cells during early differentiation is studied. The shaded area indicates the processes taking place in this time window. This figure is based on schematics of embryogenesis taken from (Nakaki and Saitou, 2014; Deuve and Avner, 2011; Takaoka and Hamada, 2012; Reik, 2007; Augui et al., 2011). For the X chromosome, Xp (Xm) indicates the paternally (maternally) inherited X chromosome.

These mechanisms will be studied using mainly transcriptome data generated with the help of microarrays. Using a densely sampled time series of the early differentiation transcriptome, I study the following two aspects. First, how X-chromosome inactivation is coupled to the exit from pluripotency and second, how gene expression during X inactivation and differentiation in general is influenced by chromatin conformation, giving rise to functional modules of co-expressed genes. To investigate how transcription factors regulate gene expression during differentiation, I will make use of a large data set of published transcriptome data from mouse embryonic stem cells.

This work is composed of three chapters, in each of which expression patterns of differentiating mouse embryonic stem cells are used to investigate a regulatory mechanism. The first chapter deals with expression on the X-chromosome. In addition to the well studied process of X chromosome inactivation in female cells, I find evidence for the previously proposed mechanism of upregulation of X-linked genes in male cells. I also study the coupling of X chromosome inactivation with the progress of differentiation, discovering a feedback of the inactivation status to differentiation.

In the second chapter I analyze expression patterns on the X chromosome during differentiation in greater detail. I observe that coordinated expression in groups of neighboring genes is linked to increased chromatin contacts of these genes' loci. Extending the analysis, I find that the alignment of co-expression domains with domains of increased chromatin contact applies genome-wide.

In the third chapter, I further increase the level of detail by investigating how well transcription factor regulation can be predicted from the expression profile of an individual gene. Using a large data set of public expression data from mouse embryonic stem cells, I compare different algorithms that reconstruct gene regulatory networks from transcriptome data. To determine the most biologically valid network, the reconstructed networks are then benchmarked with gold standards for interactions of transcription factors with genes. I find that the top performing algorithms show little difference in performance but predict fundamentally different network topologies.

All three chapters analyze regulatory processes in differentiating mouse embryonic stem cells. Central to differentiation is a network of transcription factors that both maintains the pluripotent state and controls the exit from this state (Xue et al., 2011). The exit from pluripotency triggers several downstream processes, among them inactivation of the X chromosome in female cells, spatial rearrangement of chromatin in the nucleus and a fundamental change in the activity of transcription factors. All three of these regulatory processes are controlled and possibly also control the pluripotency maintaining network of transcription factors. In the following, I will discuss what is known about these regulatory processes during differentiation. To the extent that it has been characterized, I will focus on the control of these processes by the pluripotency-maintaining factors. I will start this discussion by briefly placing embryonic stem cells into the context of embryogenesis, followed by a characterization of the pluripotency maintaining network.

1.1 Mouse embryonic stem cells: control of the pluripotent state

Embryonic stem (ES) cells form an *in vitro* system with remarkable properties. They can undergo extensive *in vitro* expansion and, upon being introduced into the early mouse embryo, can give rise to all somatic and germ line tissue (Martello and Smith, 2014). This naive, unbiased state is called pluripotency. In order to characterize embryonic stem cells, we will give a brief overview of the process of embryogenesis in the following. We will focus on the time window corresponding to the early differentiation of embryonic stem cells (Fig. 1.1).

The blastocyst forms around E3.5 (embryonic day 3.5) and is comprised of the inner cell mass surrounded by the trophoectoderm. At the late blastocyst stage (E4.5), the inner cell mass segregates into the early epiblast, that expresses the transcription factor Nanog, and the extra-embryonic primitive endoderm. The epiblast is the source of all embryonic lineages, including the germ line. Embryonic stem cells are derived from the mid-blastocyst-stage inner cell mass and most closely resemble the late blastocyst pre-implantation epiblast at E4.5 (Boroviak et al., 2014). The differentiation process studied in this work starts with embryonic stem cells that are differentiated into epiblast like cells, corresponding to the transition from the pre-implantation epiblast to the post-implantation, pre-gastrulation epiblast (Hayashi et al., 2011). The epiblast like cells have lost pluripotency and are called 'primed' because they express lineage specification factors. Several processes that impact a large fraction of the transcriptome fall into this time window, among them the marked downregulation of pluripotency associated genes, re-acquisition of DNA methylation, random X inactivation and the spatial reconfiguration of chromatin. Some of these processes will be discussed in detail in the following.

Embryonic stem cells can be propagated in the pluripotent state by adding inhibitors for certain signaling pathways to the cell culture. These signaling pathways

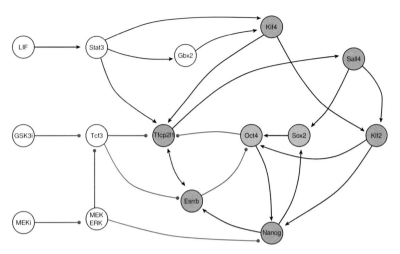

Figure 1.2: Model for the coupling of the pluripotency-maintaining network to signaling pathways. The interaction scheme is a graphical depiction of a Boolean model that is able to describe observations under a wide range of culturing conditions (Dunn et al., 2014). It contains experimentally observed as well as predicted interactions, with activating interactions in black and inhibiting interactions in red. Based on this network and independent observations, Oct4 and Sox2 (magenta) have been dubbed core pluripotency factors, while the transcription factors indicated in blue have been called ancillary factors. The ancillary factors shield the core factors from a direct interaction with extrinsic signaling pathways, presumably to safeguard the essential core factors against premature disturbance of the pluripotent state. Figure adapted from Dunn et al. (2014).

are coupled to a network of transcription factors that are essential for the maintenance of the pluripotent state (Huang et al., 2015; Hackett and Surani, 2014). The chemically well defined culture conditions used for maintaining and expanding embryonic stem cells are the result of a long process of refinement, during which key players of the pluripotency-maintaining network and their wiring were characterized (Martello and Smith, 2014). In the following we will give an overview of this network (Fig. 1.2), its coupling to upstream signaling pathways and its control over downstream processes.

Members of the pluripotency-associated network

Several transcription factors are now known to play a central role in maintaining the pluripotent state and in regulating differentiation. Among these transcription

factors, the POU-domain transcription factor Oct4 (Pou5f1) and SRY-box transcription factor Sox2 stand out because they are required for both acquisition and
maintenance of the pluripotent state. While mouse embryonic stem cells (mESC)
under standard cultivating conditions express Oct4 and Sox2 at high levels, absence of Oct4 and Sox2 triggers differentiation (Avilion et al., 2003; Nichols et al.,
1998).

In addition to Oct4 and Sox2, the transcription factor Nanog is an important
regulator of pluripotency. However, while Oct4 and Sox2 have been shown to be
essential for ESC self-renewal, Nanog expression is dispensable for renewal under
optimal culture conditions (Martello and Smith, 2014). Furthermore, in contrast
to Oct4 and Sox2 it is possible to generate induced pluripotent stem cells without
forced expression of Nanog (Schwarz et al., 2014; Carter et al., 2014). Despite
the fact that Nanog is not as closely connected with the pluripotent state as Oct4
and Sox2, the three transcription factors form a tight network, mutually regulating
each other. Also the factors Oct4, Sox2 and Nanog (OSN) have a high overlap in
genomic targets, to which they recruit further co-activators (Kim et al., 2008).

Further transcription factors have been identified that stabilize pluripotency but
are not essential for maintaining it. Because of this less critical role they have also
been dubbed ancillary factors. The genes included among these ancillary factors
varies, but Klf2, Klf4, Esrrb, Sall4, Tfcp2l1 are among the consensus members of
this group (Hackett and Surani, 2014; Kalkan and Smith, 2014).

Several of the stabilizing factors have been found to interact with genes that
destabilize the pluripotent state and facilitate progression of differentiation. Among
the most important of these destabilizers are Tcf3 and Pum1. The RNA-binding
protein Pum1 represses the ancillary factors Tfcp2l1, Esrrb, Tbx3 and Klf2 by
binding the 3' untranslated region of their mRNAs. Knocking down Pum1 leads to
a delayed downregulation of these factors during differentiation (Leeb et al., 2014).

The transcription factor Tcf3 represses Tfcp2l1, Esrrb, Nanog and Klf2. Knockout of Tcf3 leads to increased expression of these ancillary factors and slows down
the exit from pluripotency (Yi et al., 2008). The destabilizing factor Tcf3 is also
an important link to signaling activity upstream of the pluripotency gene regulatory network. The Wnt-pathway component β-catenin sequesters Tcf3, reducing
its repressive effect (Faunes et al., 2013). In addition to the Wnt-pathway, the
MAPK-pathway feeds into the pluripotency gene regulatory network.

Upstream processes triggering exit from pluripotency   The current understanding of the signaling pathways converging on the core pluripotency network was
developed along with the optimization of culture conditions of embryonic stem

cells. Initially, embryonic stem cells were cultured in fetal calf serum with added leukemia inhibitory factor (LIF). In these culture conditions, cells were found to exhibit considerable heterogeneity in expression state, with only a subset of cells being pluripotent (Graf and Stadtfeld, 2008).

Subsequent chemical refinement of culture conditions led to the development of a culture system consisting of two inhibitors, hence the name 2i medium (Ying et al., 2008). The inhibitors repress Gsk3, a part of the Wnt-pathway and Mek, part of the MAPK-pathway.In the presence of these two inhibitors, cells are found to be closer to a hypothetical ground state, corresponding to a completely naive and unbiased state (Silva and Smith, 2008; Marks et al., 2012). This is evidenced by the fact that cells cultured under these conditions are almost completely liberated from epigenetic constraints of expression. Furthermore, they contribute much more efficiently to the embryonic tissue when injected back into the blastocyst (Ying et al., 2008). The direct targets of the Wnt-pathway and the MAPK-pathway in the core network were found to be Tfcp2l1, Klf2, Esrrb and Nanog (Martello and Smith, 2014).

A Boolean model integrating observations from many different culture conditions posits that with respect to signaling, the pluripotency-maintaining network is constructed in a layered fashion (Dunn et al., 2014) (Fig. 1.2). The core, consisting of Oct4 and Sox2, directly interacts with the ancillary factors, among which are Tfcp2l1, Klf2, Esrrb and Nanog. These factors then in turn form the interface to external signaling events. In this way, the core is shielded from direct external influence, possibly to ensure a certain degree of robustness of the decision to differentiate.

1.2 Downstream differentiation events triggered by the pluripotency network

The destabilization of the pluripotency maintaining network triggers a sequence of events leading to the irreversible commitment of cells to different lineages that ultimately give rise to somatic tissue. How precisely pluripotency associated transcription factors trigger these events is still poorly understood. A better understanding of the control of pluripotency factors over downstream processes promises insights into the remarkable properties of differentiation.

Proper cellular differentiation requires both synchronous regulation of genes with overlapping tasks as well as temporal compartmentalization of genes with different tasks. This may require feedbacks that make sure that certain processes are kept on hold until others have progressed sufficiently. An additional layer of complexity stems from the need for tissue specific expression. This requires some modularity of gene regulation so that genes needed for a specific tissue can be controlled together. On top of this, gene-specific cis-regulatory elements ensure the individual tuning of gene expression.

Gene regulatory changes triggered by differentiation   After withdrawal of 2i, hundreds of genes are expressed *de novo* within the next 48 h. Together with the downregulation of pluripotency-associated genes, this amounts to a global reorganization of the transcriptome (Buecker et al., 2014). What is the precise sequence of these reorganization events and how are they triggered by the exit from pluripotency? In contrast to considerable advances in characterizing the pluripotency-maintaining network, the links of this network to downstream processes are not well understood.

Important insights about the role of the reprogramming factors Oct4, Sox2 and Nanog came from early ChIP-seq experiments. One of the most important observations made at that time was that the majority of Oct4 binding sites are also bound by Sox2 (Loh et al., 2006). This so-called combinatorial binding was confirmed by more extended studies of binding sites of pluripotency-associated factors. A later study found that around 800 promoters are bound by four or more pluripotency-associated factors (Kim et al., 2008). Also, the authors observed a correlation between the number of factors bound at a promoter and expression of the respective gene in ES cells.

A subsequent statistical analysis could show that a model using binding of 12 pluripotency-associated factors as input could explain a large fraction of the variance in gene expression in ES cells (Ouyang et al., 2009). Additional support for the importance of combinatorial binding came from a study that looked at the conservation of binding sites in ES cells between human and mouse (Göke et al., 2011). In contrast to individual binding sites, which differed between the two organisms, combinatorial binding sites were found to be conserved.

An important complimentary source of transcription factor targets are high-throughput knockdown or overexpression screens. A particularly large screen was performed by the group of Minoru Ko (Nishiyama et al., 2013; Correa-Cerro et al., 2011; Nishiyama et al., 2009). These studies confirmed the central role of Oct4 and Sox2, who induced huge perturbations in the transcriptome when overexpressed. The widespread perturbations caused by Oct4, Sox2 and Klf4 overexpression are surprising since they are already highly expressed in undifferentiated ES cells.

In addition to high-throughput experiments, literature mining has also been used to assemble the gene regulatory network of pluripotent stem cells (Som et al., 2010). The resulting network indicates an absolutely dominating role of Oct4, Sox2 and Nanog as network hubs. While the central role is supported by high-throughput experiments, the interest in the reprogramming factors might still have led to a bias in the published literature which affects the literature-derived network. Despite the possible bias in the overall structure, this network is very useful because individual links are verified, high-confidence interactions.

Finally, with the accumulation of publicly available transcriptome data, it became possible to use reconstruction algorithms that infer interactions from co-expression. An example for this approach is the study by Cegli and coworkers who applied the ARACNE algorithm on around 200 microarray samples (Cegli et al., 2013). The reconstructed network was restricted to genes that are expressed in mouse ES cells to obtain a regulatory network that is specific for these cells. The hub genes of the reconstructed network were used to select a previously uncharacterized gene for a follow-up analysis.

With all the different approaches available, the question emerges which of these approaches are the most useful ones. All approaches of course suffer from drawbacks. Networks based on co-expression are not directional, in contrast to networks based on ChIP-seq data or TF knockdown. Binding sites identified by ChIP-seq identify direct interaction of the transcription factor with nearby genes but not all binding sites are functional. Differential expression upon TF knockdown is an important criterion for a functional interaction but it does not necessarily have to be direct. Currently all high-throughput methods are limited by the number of TFs that can be interrogated. These limitations do not apply to network reconstruction algorithms, which however are subject to limitations beyond directionality (Margolin and Califano, 2007). The precision of the predictions is typically rather low. The reason lies in the difficulty to separate indirect from direct interactions, for which a number of corrections have been developed.

With all approaches suffering from serious limitations, an important goal at the moment has to be to find out how well different methods agree on interactions to arrive at a consensus network supported by multiple types of evidence. Only after progress has been made in this direction can large scale gene regulatory networks be used to make predictions on a systemic level. In chapter 4 of this work, the overlap of interactions predicted by different network reconstruction algorithms with high-throughput transcription factor perturbation and binding data will be studied.

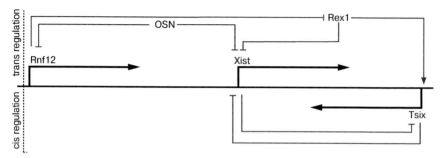

Figure 1.3: Central genes involved in the control of Xist. The schematic illustrates the location of three X-linked genes, the protein-coding gene Rnf12 and the non-coding genes Xist and its anti-sense repressor Tsix. Cis-acting regulation is indicated in the lower part of the figure, trans-acting regulation in the upper part. Important trans-acting factors include OSN (Oct4, Sox2 and Nanog) as well as the pluripotency-associated factor Rex1 (Zfp42). The presented network is a simplified version of the scheme published in Dupont and Gribnau (2013).

X inactivation control by pluripotency factors Inactivation of the X chromosome in female embryos is needed to equalize the gene dosage of X-linked genes between female and male cells (Minkovsky et al., 2012). Around embryonic day 5.5 either the maternal or paternal X chromosome in females is randomly inactivated in the epiblast lineage. The fact that this process can be recapitulated in embryonic stem cells made it possible to study the control of random X inactivation in detail (Lee, 2011) (see also Fig. 1.1, page 2).

One of the earliest insights in the genetic control of X inactivation was the discovery of the X-linked region needed for X inactivation: the X inactivation center (Rastan, 1983). In the following we will briefly describe how X inactivation is regulated by an interplay of the core pluripotency factors with genes located in the X inactivation center.

The inactivation of the X chromosome is mediated by monoallelic expression of a non-coding RNA called Xist, that is located in the X inactivation center. Studies indicate that Xist coats the chromosome it is expressed from (Herzing et al., 1997). Control of this non-coding RNA is thus central to control over the process of X inactivation. Consistent with its central role, Xist is subject to both trans-acting regulation from pluripotency-associated factors and trans- as well as cis-acting regulation by genes located in the X inactivation center (Augui et al., 2011) (Fig. 1.3).

The regulation by the pluripotency-associated factors ensures the coupling of differentiation and X inactivation. Additionally, trans- and cis-acting X-linked genes and regulatory elements are needed two implement both the master switch that triggers X inactivation and an X chromosome counting mechanism (Augui et al., 2011). The latter is needed to restrict X inactivation to female cells, i.e. cells with two X chromosomes. The most important X-linked genes regulating Xist expression are Rnf12 and Tsix. The regulatory interplay of these genes with Xist and the pluripotency factors lies at the heart of the X inactivation process.

Regulation of X inactivation through the pluripotency-maintaining network is mediated by some of its central members, among them Oct4, Sox2 and Nanog, that all bind within the first intron of Xist. Though there are doubts that this is a functional binding site, proof for at least indirect repression comes from knockout experiments, showing that loss of Nanog or Oct4 leads to Xist upregulation in differentiating male ESCs (Navarro et al., 2008). Additionally, Oct4, Sox2 and Nanog repress the protein-coding, X-linked gene Rnf12 (Navarro et al., 2011). The gene Rnf12 codes for a trans-acting factor, which acts as an activator of Xist. Though not essential for Xist expression, overexpression of Rnf12 can induce Xist and subsequent X inactivation in male ESCs (Augui et al., 2011). The Xist activator Rnf12 acts by degrading the Xist repressor Rex1 (Gontan et al., 2012).

The anti-sense transcription unit of Xist harbors the Tsix gene, which represses Xist. This was shown by deleting regions 3' to Xist which leads to non-random Xist upregulation and X inactivation (Clerc and Avner, 1998). How Tsix prevents Xist upregulation is the subject of ongoing research. Several reports suggest that Tsix transcription silences Xist by mediating chromatin modifications (Navarro et al., 2005; Sado et al., 2005). In undifferentiated cells, the Xist promoter is enriched in repressive marks, while the Tsix promoter is enriched in euchromatin-associated histone marks. This changes in female differentiating cells, where the Xist promoter shifts to a euchromatic state and the Tsix promoter acquires repressive marks (Navarro et al., 2009). The spread of these marks seems to be controlled by Tsix, since its deletion leads to a loss of repressive marks at the Xist promoter in undifferentiated cells (Navarro et al., 2006).

This fact points to the role of Tsix gene or regulatory elements contained in this gene for the organization of chromatin at the X inactivation center. This organization is connected with the presence of binding sites of the CTCF protein, that is known to insulate regions of the genome with respect to chromatin modifications, downstream of the transcription start site of Xist. It has been proposed that these binding sites separate the Xist 5' region and the Tsix transcriptional unit into two distinct domains with opposing Tsix action (Navarro and Avner, 2010).

In addition to the genes described above, multiple cis-regulatory elements which modulate the expression of these genes have been described. The emerging picture of the regulation of X inactivation is thus a very complex one. Consequently, it is not surprising that open questions remain. Unknown regulatory elements or mechanisms could explain why even the largest transgenes can not recapitulate proper Xist expression (Heard et al., 1999). Another open question is whether the control of pluripotency-associated factors over the X inactivation center is complemented by a feedback of X-linked genes on these factors. These questions will be addressed in Chapters 2 and 3 of this work.

Control of gene expression through localization in the nucleus In addition to DNA-binding proteins and epigenetic regulation, expression can be controlled by the spatial localization of chromatin in the nucleus. Chromatin can be classified into two different classes based on localization and transcriptional activity, heterochromatin and euchromatin. Heterochromatin contains mostly silenced genes and is localized at the nuclear periphery and near the nucleolus. Euchromatin corresponds to transcriptionally active genes and is situated between heterochromatin domains (Joffe et al., 2010).

Using chromosome conformation capture, Lieberman-Aiden and coworkers explored the spatial organization of chromatin in greater detail (Lieberman-Aiden et al., 2009). They found chromatin to aggregate in two distinct compartments, labeled A and B, at scales ranging from tens to hundreds of megabases. Since compartment A is associated with actively transcribed, gene rich regions it was identified with open, accessible chromatin. Compartment B corresponds to closed, inaccessible chromatin. This spatial segregation of open, active chromatin from closed, inactive chromatin shows that inter-loci interactions are connected with similar expression. Thus, the spatial organization of chromatin clearly has functional implications.

For the connection between expression changes during differentiation and spatial organization of chromatin, association of chromatin with the nuclear lamina proved important. The localization of heterochromatin to the nuclear periphery is aided by interaction with the nuclear lamina. In drosophila, mouse and human, domains of genes associated with the lamina have been found. These so-called lamina associated domains (LADs) are transcriptionally repressed. A study by Peric-Hupkes et al. (2010) found that LADs can be bound or released from the lamina during differentiation. These authors compared the LAD organization in embryonic stem cells to that of neural precursors cells and terminally differentiated astrocytes. They found the biggest difference in LAD structure between ESC and astrocytes, whereas differences between neural progenitors and astrocytes were smaller. Additionally, changes in LAD structure were associated with changes in expression. Binding of LADs to the lamina and release from the lamina were associated with repression and activation, respectively.

Another important example for the functional relevance of changes in spatial organization of chromatin comes from a study of looking at the Hox gene cluster (Morey et al., 2009). In this study, the localization of Hox gene alleles with respect to chromosome territories was studied. The authors showed that transcribed Hoxb and Hoxd alleles relocate to the border of chromosome territories during differentiation, whereas silent alleles remain in the interior of these territories. Together with the investigation of LAD dynamics, these observations show that chromatin is spatially reorganized during differentiation. This reorganization is important for activation and repression of genes as cells progress trough differentiation stages.

Despite these recent advances in understanding how spatial reorganization impacts gene expression, several questions remain open. One of these questions is whether the isolated observations on the impact of spatial localization on gene expression are examples of a general regulatory principle. It remains furthermore unclear if spatial domains are generally related to functional groups of genes as observed in the Hox cluster. These questions will be addressed in chapter 3 of this work.

2 Global regulation of expression in early differentiation of mESCs

2.1 The X:A ratio is fine tuned by two opposing processes

The gene expression data for differentiating XO, XY and XX mESC used in this sub-chapter has been generated by Edda G. Schulz.

2.1.1 Introduction

In mammals, sex is determined by sex chromosome complement, XY in males and XX in females. In males, most X-linked genes are present only in one copy, leading to a difference in the ratio of X-linked genes and autosomal genes (X:A ratio) between the sexes. For this reason, compensatory mechanisms have evolved that balance the X:A ratio between males and females. One of these processes is the random X inactivation of one X copy in females. While this process is well studied, the opposing mechanism of upregulation of the single active X remains less well characterized and more controversial in extent. We compared the impact of both processes over the course of differentiation of mESCs and found that the extent of X upregulation depends on the evolutionary history of a gene. We also show that during differentiation compensatory mechanisms lead to a balancing of the X dosage between the sexes as well as balancing of X:A ratios for each sex.

The present day mammalian X and Y chromosomes developed from the proto-X autosome of the common ancestor of mammals and birds. While therian mammals developed the XY sex chromosome system, birds independently developed the ZW system where females are heterogametic (Smith et al., 2007). In mammals, one copy of the X chromosome lost most of its genes in the process of becoming the modern Y chromosome. How could this deterioration of the Y chromosome be tolerated when autosomal monosomy is lethal during mammalian development?

To resolve this conundrum, Ohno proposed that the single active X chromosome must have acquired an upregulation mechanism during the course of evolution (Ohno, 1967). Together with X inactivation in females, this mechanism should ensure a one-to-one expression ratio between X-linked and autosomal genes in both sexes. With the advent of high-throughput transcriptome analysis it has become possible to test Ohno's hypothesis in a chromosome-wide manner. However, far from firmly establishing Ohno's hypothesis, this has led to a controversy in the literature with claims ranging from complete compensatory upregulation to no upregulation at all. An overview of publications supporting or refuting Ohno's hypothesis is given in Tab. 2.1.

The global upregulation hypothesis, stating that X-linked alleles are expressed on average at the twofold level of autosomal alleles, was confirmed by early microarray based studies. Nguyen and Disteche showed that $X : A \approx 1$ in human and mouse somatic tissue as well as mouse embryonic tissue (Nguyen and Disteche, 2006). These results were confirmed by Lin et al. (2007), who detected a global upregulation of X-linked genes relative to autosomal genes in mouse embryonic stem cells. This study also found an upregulation of X-linked genes over progressive stages of stem cell differentiation.

Later an RNA-seq-based study (Xiong et al., 2010) claimed that no dosage compensation can be detected and the observed X:A ratios are compatible with 0.5. Comparing microarrays and RNA-seq, the authors proposed that initial studies failed to detect a lack of dosage compensation because microarrays compress differences in expression.

At this stage of the dispute, three hypotheses were proposed that modify Ohno's hypothesis in the sense that it should apply only to a relevant subset of X-linked genes. Depending on whether this subset forms the majority or minority of the X-linked genes, the respective authors supported or questioned Ohno's hypothesis.

One subset of genes that was proposed to be compensated are dosage sensitive or haploinsufficient genes. Proponents of this hypothesis argue that for many genes, cells can easily cope with dosage fluctuations. Analyses of dosage compensation should thus focus on groups of genes where there is a strong indication that they are dosage sensitive (Mank et al., 2011; Pessia et al., 2014). Indeed, it was shown that X-linked genes involved in large complexes with more than 7 proteins are dosage compensated and that their dosage increases with the complex size (Pessia et al., 2012; Lin et al., 2012). Because of the small size of this gene set, these authors refuted Ohno's hypothesis.

Table 2.1: Publications supporting or refuting Ohno's hypothesis, gathered in Pessia et al. (2014). The column 'Ohno' indicates whether the publication supports Ohno's hypothesis or not.

Publication	Ohno	Gene filtering	Technology	Tissue/Cells
Nguyen and Disteche (2006)	yes	lowly expressed	microarray	somatic human, mouse and mouse embryonic
Deng et al. (2011)	yes	lowly expressed	RNA-seq	somatic human and mouse
Kharchenko et al. (2011)	yes	lowly expressed	RNA-seq	somatic human and mouse tissue
Lin et al. (2011)	yes	implicit by platform	microarray	mouse embryonic stem cells
Yildirim et al. (2012)	yes	lowly expressed	RNA-seq	mouse embryonic kidney fibroblasts
Xiong et al. (2010)	no	all mapped genes	RNA-seq	mouse liver, brain, muscle
He et al. (2011)	no	all mapped genes	RNA-seq	human heart
Castagné et al. (2011)	no	compare filters	microarray	human monocytes, mouse heart

The second partial compensation hypothesis states that Ohno's hypothesis applies only to X-linked genes that were already present on the proto-X-chromosome of the common ancestor of mammals and birds. These genes are called ancestral genes, while the rest of the genes are called acquired genes. This hypothesis is derived from the argument that only those X-linked genes present during the evolutionary sex chromosome differentiation would suffer from a dosage imbalance with respect to the autosomes (He et al., 2011; Julien et al., 2012). Multiple studies, however, found no upregulation of ancestral X-linked genes (Lin et al., 2012; He et al., 2011; Julien et al., 2012). Contrary to these results, Deng et al. (2013) found that both ancestral and acquired X-linked genes are upregulated. Additionally, these authors found some evidence for a stronger upregulation of ancestral genes by the histone acetyltransferase MOF.

Noting that the X chromosome harbors many testis-specific genes that are not expressed in other tissue, proponents of the third partial compensation hypothesis argued that lowly expressed X-linked genes should be excluded from the analysis. Supporting this hypothesis, Deng et al. (2011) found the X chromosome to be upregulated when excluding these genes. They also observed an increased occupancy of the RNA polymerase II at highly expressed X-linked genes compared to highly expressed autosomal genes.

There is presently no agreement over the existence or extent of upregulation of X-linked genes. One compounding problem for a comparison of the results is that each study differs in some parameters, among them the filtering of genes, the selection of tissue studied and the platform used for transcriptome analysis. Filtering of genes has emerged as the main point of criticism voiced by the opponents of the global upregulation hypothesis (Castagné et al., 2011).

Two kinds of filtering can be identified in the literature, the implicit filtering of genes by the platform used to assay the transcriptome and explicit filtering of lowly expressed genes. Implicit filtering in microarrays can be caused by a biased coverage of the genome by the probes employed or failure to separate lowly expressed genes from the background noise, while for RNA-seq it is mainly a consequence of the employed sequencing depth. A reduced fraction of X-linked genes compared to autosomal genes has been reported for an older microarray type (Lin et al., 2011). Explicit filtering relies on thresholds or strategies that may be hard to justify. One study compared two different filtering strategies (Castagné et al., 2011). The first strategy used an absolute expression threshold applied to all chromosomes, while the second used filtering out a fixed proportion of genes for each chromosome. The study indicated that X:A ratios differ considerably between the two strategies.

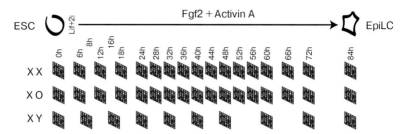

Figure 2.1: Experimental setup and timing of measurements: Three mESC cell lines, XX (PGK12.1), XO (derived from PGK12.1) and XY (E14) cells were cultured in serum free 2i medium and LIF. Starting at experiment time 0 h cells were differentiated for 3.5 days into epiblast like cells (EpiLC) using Fgf2 and Activin A. We harvested RNA at the indicated time points and assessed mRNA expression using Affymetrix mouse exon arrays.

In addition to implicit filtering, it has been shown that microarrays compress expression differences for lowly expressed genes (Compare Perkins et al. (2014) and references therein). Though RNA-seq is considered to be superior to microarrays, it has also been shown that the X:A ratio derived from sequencing data is highly dependent on sequencing depth, read mapping and other evaluation pipeline parameters (Jue et al., 2013).

Because of the strong influence of gene filtering and the technical limitations of the platform used, we focus on the change of gene dosage instead of trying to precisely quantify gene dosage in a given tissue or stage of differentiation. We use a time course of differentiating mouse embryonic stem cells to find signs of gene dosage compensation. Tracking ancestral and acquired genes over time we show that ancestral X-linked genes are upregulated relative to autosomal X-linked genes. For acquired genes, we find a higher X:A ratio but no upregulation over time. These microarray-based results are supported by our analysis of published RNA-seq data. We also discuss the impact of implicit and explicit filtering on our results. Further studies should clarify whether this upregulation is transient or whether it is maintained in somatic tissue.

2.1.2 The transcriptome time series captures the signature of X inactivation

The experimental system employed in this study are mESCs during early differentiation. We obtained a densely sampled transcriptome time course during the first 3 days of differentiation of mESC into epiblast like cells (EpiLC) for three cell lines with distinct sex chromosome complement: a female XX line (Pgk12.1), an XO sub-clone of this line that lost one X chromosome and a male XY line of related genetic background (E14), Fig. 2.1. This *in vitro* system is suited to studying X inactivation and X dosage compensation more generally, because it models the transition from the early, pre-implantation epiblast to the post-implantation, pre-gastrulation epiblast (Hayashi et al., 2011). Around the time point of implantation, random X inactivation occurs in the mouse embryo (See Fig. 1.1, page 2).

As discussed in the introduction to this chapter, one of the problems in assessing expression ratios between chromosomes comes from implicit filtering due to the method with which the transcriptome is assayed. For microarrays, probe design and the distribution of genes falling under the detection threshold among the chromosomes lead to this implicit filtering. We can estimate the extent of implicit filtering by comparing the number of genes detected for each chromosome to the number of genes annotated in the Ensembl database, version 79.

When plotting the number of genes detected vs. the number of genes annotated, we found for most chromosomes that the number of detected genes is proportional to the number of annotated genes. This fact holds irrespective of whether the genes were protein coding or not (Fig. 2.2). However, the scatter around the line of perfect proportionality is considerably reduced when restricting the analysis to protein coding genes. This applies also to X-linked genes. However, for both arbitrary genes and only protein coding genes, chromosome X shows a smaller fraction of detected genes than the autosomes. Despite the remaining under-representation of X-linked genes, we performed all analyses of chromosome ratios only on protein coding genes because the variability of chromosome coverage is strongly reduced for these genes.

We next performed two checks which form a prerequisite for the following analysis. First, we checked whether our microarray samples can resolve the double dosage of X-linked genes between XX and XO/XY cells. Second, we investigated whether we can detect signs of X inactivation, so that our *in vitro* system indeed models the X inactivation process in the female mouse embryo. In order to perform both checks, we binned genes into 10 deciles according to their mean expression in the three cell lines in undifferentiated stem cells. Then we computed the average log2 ratio in expression between the XX cell line and the XO (XX:XO ratio) or XY (XX:XY ratio) cell line (Fig. 2.3).

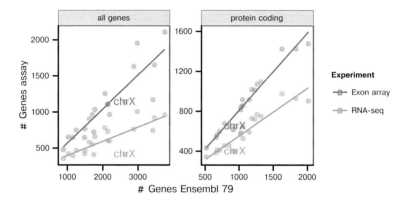

Figure 2.2: Comparison of the genes detected by the Affymetrix Mouse Exon Array (time series of XX, XO and XY cell lines) and the genes detected by RNA-seq (male and female ESCs vs. male and female NPCs, discussed in section 2.1.4). For each chromosome, the number of detected genes (vertical axis) was compared to the number of genes annotated in the Ensembl annotation, Version 79 for arbitrary genes (left panel) and protein coding genes only (right panel). The lines indicate a linear fit to the data. For the microarrays, present genes were determined according to Sec. 6.1, page 129, with the exception that pseudogenes were retained. For the RNA-seq data set, the genes for which expression was supplied in the dataset of Gendrel et al. (2014) were used.

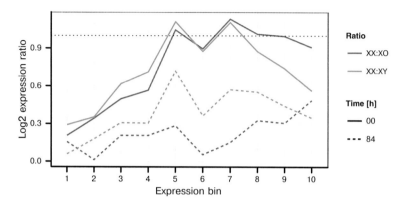

Figure 2.3: For X-linked genes, expression ratio between cell lines depends on X chromosome complement. Genes are sorted into bins corresponding to their mean expression in undifferentiated XX, XO and XY ESC. The average expression ratio between cell lines of the indicated pair is then plotted against the expression bin. Above median expression, the ratio in undifferentiated mESC (d0) is consistent with an XX:X ratio of two (1 in log2 units). Note the failure of the microarrays employed to resolve the expression differences for lowly expressed genes.

We observed that both ratios increase towards bins of higher expression in undifferentiated stem cells. This is an indication of the failure of microarrays to resolve differences in expression for lowly expressed genes, leading to a compression of such differences. For highly expressed genes, however, we observe a ratio that is compatible with the expected ratio of 1 logarithmic unit, corresponding to a linear fold change of two. This dosage difference is clearly compensated over time, to a level of a linear fold change less than approximately 1.25 for the XX:XO ratio.

2.1.3 No X chromosome upregulation in undifferentiated cells

As a first step in the analysis of dosage compensation during differentiation, we assessed possible dosage differences between chromosomes in undifferentiated stem cells in an unbiased way. Our analysis departs from the simple null hypothesis that each chromosome has the same distribution of expression as all autosomal genes. Since we observed the microarrays to underestimate expression differences for lowly expressed genes, we filtered these genes out. To define the set of lowly expressed genes to be excluded, we required a gene to be expressed above the median in undifferentiated XX cells. We chose the XX cell line for filtering since under our null hypothesis using the XO or XY cell lines for filtering should lead to a bias: The presence of only one allele for each X-linked gene would lead to a biased filtering of genes that would exclude a higher proportion of X-linked genes than autosomal genes.

Using the remaining highly expressed genes we tested the null hypothesis. The distribution of the gene expression for each individual chromosome relative to the autosomal expression is shown in Fig. 2.4a. We noticed that the X chromosome shows the strongest deviation from the average autosomal expression. Some deviation is also found for chromosomes 11, 8 and 7. This is reflected in the p-values of a two-sided Wilcoxon rank-sum test. Apart from the X chromosome, we obtained significant deviations for chromosome 11 in XO cells and for chromosome 8 in XX cells (Fig. 2.4b). Chromosome 7 shows borderline significance. Consistent with the observed excess dosage, the XX cell line (Pgk12.1) is known to harbor a trisomy of chromosome 8. As trisomies in chromosome 8 and 11 are among the most common chromosomal aberrations in mESCs, the excess dosage of chromosome 11 in the XO cell line may also stem from a trisomy (D'Hulst et al., 2013; Kim et al., 2013).

In the dispute over Ohno's hypothesis it has been pointed out that filtering of lowly expressed genes amounts to implicit assumptions about the evolutionary history of X-linked genes (He et al., 2011) and that such a filtering may exclude a higher fraction of X-linked than autosomal genes (Castagné et al., 2011). We checked this by comparing the average ratio of expressed genes to annotated genes (Ensembl v79) for all autosomes to the ratio for X-linked genes. For the autosomes, we find a ratio of $41.4\pm 4.4\%$, while for the X chromosome, we find a ratio of 36%. Thus the fraction of X-linked genes is a the lower end of the ratios found for the autosomes, but within the fluctuation of the autosomal ratios.

To address remaining concerns, we computed the ratio of median X-linked expression to median autosomal expression for different thresholds. While the resulting ratios vary with the threshold, especially for low thresholds, we consistently observed a higher X-linked than autosomal expression in the XX cell line and a lower expression in the XO and XY cell lines (Fig. 2.5).

Taken together, the analysis of the relative X dosage in undifferentiated stem cells shows that the expression of X-linked genes is significantly lower than that of autosomal genes in the XO and XY cell lines. Consequently we can exclude the possibility of dosage compensation at this stage of the differentiation process.

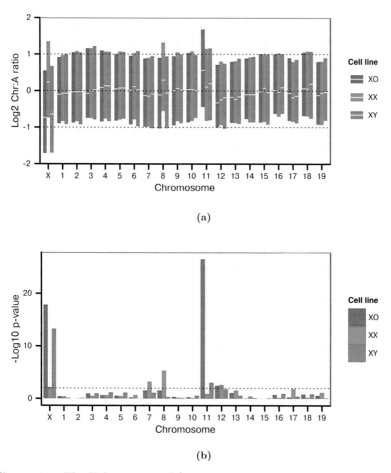

(a)

(b)

Figure 2.4: The X:A ratio in XO/XY cells is significantly lower than one. (a) Distribution of median gene expression on individual chromosomes relative to the median of autosomal expression in undifferentiated mESC. Only protein coding genes with expression above the median in undifferentiated XX cells are taken into account. The filled bars indicate the interquartile range and the median is marked by a horizontal white line. (b) P values (-log10) for two sided Wilcoxon rank-sum tests comparing the expression of genes from the indicated chromosome to the expression of all autosomal genes.

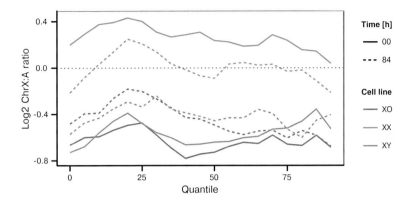

Figure 2.5: Relative dosage of the X chromosome to the autosomes does not depend strongly on expression cutoff. The expression ratio of X-linked genes to the median expression of all autosomal genes is plotted for gene filters of different stringency. Quantiles indicate the percentage of genes falling under the expression threshold.

2.1.4 Ancestral X-linked genes are upregulated during differentiation

We next asked whether the lack of dosage compensation observed in undifferentiated male stem cells is maintained during differentiation or if we can detect an upregulation of X-linked genes. Evidence for such an upregulation during differentiation has been observed before in mESC (Lin et al., 2007), though for this study a microarray was used that could only detect 252 X-linked genes (out of 940 protein coding genes annotated in Ensembl v79), indicating a possible bias to highly expressed genes. Some indication for an upregulation of X-linked genes was also found in pooled male and female mouse embryonic tissue from the blastocyst stage onward (Nguyen and Disteche, 2006). Our densely sampled time series of differentiating mESC is well suited to investigate a possible upregulation of X-linked genes in detail.

We next investigated the time dependence of the X:A ratio in order to check whether there is evidence for upregulation of the X chromosome. When plotting the log2 X:A ratio for the three cell lines, we observed downregulation in the XX cell line, consistent with random X inactivation taking place (Fig. 2.6a). However, the X:A ratio in the XO and XY cell lines does not show a clear trend, raising

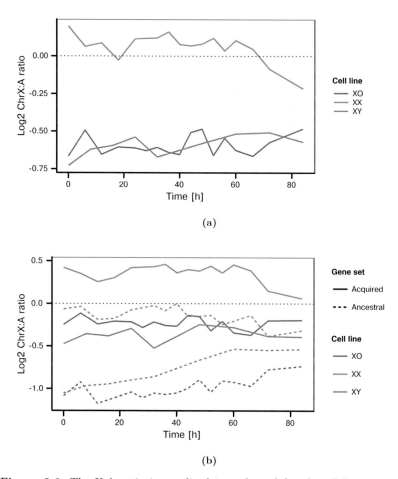

(a)

(b)

Figure 2.6: The X:A ratio is equalized in male and female cell lines over time by X inactivation and probably X chromosome upregulation. (a) Ratio of median expression of X-linked genes to median expression of all autosomal genes for the three cell lines, XO, XX and XY. (b) Ratio of median expression of ancestral X-linked genes to median expression of ancestral autosomal genes (dashed line) compared to median log2 X:A ratio of newly acquired genes (solid line).

the possibility that the observed dosage compensation between the sexes is caused exclusively by X inactivation.

As discussed in Sec. 2.1.1, X upregulation in order to balance the X:A ratio has been hypothesized to only take place for subsets of X-linked genes. In particular, it has been argued that only genes already present on the X chromosome before it differentiated into a sex chromosome should need an upregulation mechanism in order to restore the original dosage (He et al., 2011; Julien et al., 2012).

In order to test this hypothesis, we determined the genes that were present on the proto-X-chromosome (ancestral genes) and those that were acquired later (acquired genes). By proto-X-chromosome we denote the X chromosome present in the last common ancestor of birds and mammals. Ancestral genes were determined by homology of mouse genes with chicken genes (For details, see Materials and Methods, Sec. 6.2, page 129). Among the 21836 protein coding genes annotated in Ensembl v79, we identified 261 ancestral X-linked genes, 679 acquired X-linked genes, 7939 ancestral autosomal genes and 12957 acquired autosomal genes. For an overview of ancestral and acquired genes detected by the Exon arrays, see Tab. 6.1, Materials and Methods, Sec. 6.2, page 130.

We then separately computed the median ratio of ancestral (acquired) X-linked and ancestral (acquired) autosomal expression for each time point of the differentiation process (Fig. 2.6b). We noticed that for acquired genes, the X:A ratio is much higher in undifferentiated stem cells than for ancestral genes. Consistent with the hypothesis that only ancestral genes should need dosage compensation, we observed that the X:A ratio increased in both the XO and XY cell lines over time.

The cumulated density function for the ancestral genes of each chromosome shows that a consistent upregulation can only be detected on the X chromosome (Fig. 7.2, page 144 and Fig. 7.1, page 143). It should be noted that in principle the quantile normalization of microarray samples makes it impossible to discern an upregulation on the X chromosome from a uniform downregulation of all autosomes. However, such a uniform synchronized downregulation of all autosomal genes is highly unlikely. Thus, the rise in the X:A ratio is due to an upregulation of ancestral X-linked genes.

We next investigated the change in the log2 X:A ratio for ancestral and acquired genes in groups of genes defined by average expression. Because of the non-linearity of microarrays in the lower expression range and different fractions of silent genes on the X and on the autosomes, we used an approach suited to comparing entire distributions of expression values (He et al., 2011; Deng et al., 2011). For this

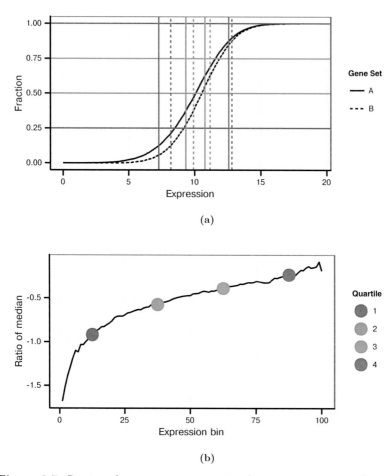

(a)

(b)

Figure 2.7: Strategy for comparing expression between two groups of genes (A, B) with distinct expression distribution in different expression bins. (a) The cumulated density function of the expression of genes in group A (solid line) and group B (dashed line) is shown. To compare expression between gene group A and gene group B, the median expression is separately computed for genes falling into different expression bins. We indicated quartiles by solid horizontal lines and the corresponding median of groups A and B by a vertical solid line and a vertical dashed line, respectively. (b) The ratio of the median expression (assuming a log scale for expression values, this is computed by taking the difference of the medians) in group A and the median expression in group B is computed for each percentile (solid line). Colored dots indicate the ratio of medians for the corresponding quartile.

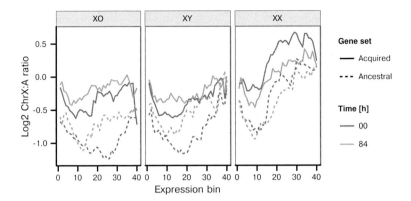

Figure 2.8: Resolving the distinct compensation behavior for different gene sets is limited by the linear range of microarrays. Applying the analysis of Fig. 2.7, genes are grouped into four categories according to evolutionary history (ancestral or newly acquired) and chromosome origin (X-linked or autosomal) and separately binned into 40 expression bins according to expression within each group and cell line. Then we calculated the ratio of the medians of X-linked and autosomal expression in each bin for ancestral (dashed line) and newly acquired (solid line) genes.

approach, genes from two groups to be compared are separately binned into equally sized expression bins. Then, for each bin the ratio of the expression median of the first group and the expression median of the second group is computed. This computation is illustrated in a conceptual figure (Fig. 2.7), comparing two sets of genes with equally distributed expression except for an excess of lowly expressed genes in set A.

Applying this method to the X:A ratio we can determine the expression range over which ancestral genes are upregulated with respect to autosomal expression. The comparison of the log2 X:A ratio for ancestral and acquired genes for the three cell lines XO, XY and XX is shown in Fig. 2.8. For undifferentiated cells of all three cell lines, the X:A ratio for acquired genes is higher across the entire expression range than the X:A ratio for ancestral genes. Over the course of differentiation, the distinct behavior of ancestral and acquired genes is confirmed. While ancestral genes in the XO and XY lines are upregulated over time especially in the medium to high expression range, little or no upregulation could be observed for acquired genes.

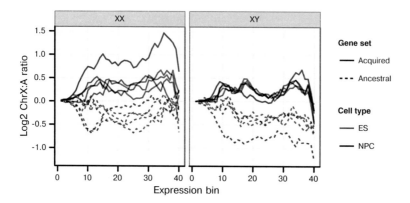

Figure 2.9: Same analysis as in Fig. 2.8 with RNA-seq data from (Gendrel et al., 2014), comparing embryonic stem cells (ES) to neural progenitor cells (NPC). Note that the RNA-seq data exhibits less sensitivity of the ChrX:A ratio with respect to the expression interval because of the extended linear detection range compared to microarrays. Note that the X:A ratio is compressed for the lowest expression bins due to the transformation of FPKM values according to log2(FPKM+1).

Consistent with a non-linear response of microarrays for lowly expressed genes, differences in expression decrease when going to very low expression bins. Despite this compression, there still is evidence for a lack of upregulation for the lower quartile of ancestral genes. This could in principle be due to an excess of silent, ancestral genes on the X chromosome that remain silent during differentiation. In addition to the observations in the XO and XY cell lines, we also observed that downregulation of ancestral genes in the XX line is strongly attenuated compared to acquired genes. This may be an effect of X inactivation. It is possible that upregulation and X inactivation effects cancel out for ancestral genes in female stem cells.

To analyze the impact of the dynamic range and to make sure that the observed upregulation effect can be confirmed by independent experiments, we contrast our results with RNA-seq data from a published study (Gendrel et al., 2014) (See Materials and Methods, Sec. 6.2, page 130). This study compared expression of male and female mouse embryonic stem cells to male and female neuronal progenitor cells. These neuronal progenitor cells correspond to a much later stage in cell differentiation than the end point (day 3) of the data discussed so far (Conti et al., 2005).

We repeated the computation of the expression bin dependent log2 X:A ratio for ancestral and acquired genes (Fig. 2.9). The RNA-seq data confirmed the observation that ancestral X-linked genes in male cells get upregulated during differentiation. In contrast to the microarray data we noticed an increase of the X:A ratio for ancestral genes in male cells over the entire expression range. We also again observed less downregulation of ancestral genes compared to acquired genes in female cells. Overall, the RNA-seq data supports the existence of a mechanism upregulating ancestral X-linked genes with respect to ancestral autosomal genes in males, possibly also in females.

It has to be noted that the under-representation of X-linked genes is worse in the RNA-seq data set than for the Exon-array-based time series, especially for arbitrary genes (see Fig. 2.2, page 21). For arbitrary genes, we find approximately half the fraction of detected over annotated genes for X-linked genes compared to the autosomes (chromosome X: 0.16, autosomes 0.32 ± 0.06). This detection problem is partially corrected for protein-coding genes, with a detection ratio of 0.42 for chromosome X and 0.56 ± 0.06 for the autosomes. This fact indicates that filtering on protein coding genes is necessary to limit the impact of implicit filtering.

2.1.5 Discussion and outlook

The X:A ratios we observe for ancestral and newly acquired genes in male and female cells are remarkably consistent with a simple model of dosage compensation. In female undifferentiated stem cells, newly acquired genes are present at a linear X:A ratio of two, while ancestral genes exhibit a linear X:A ratio of one. In male undifferentiated stem cells we find an X:A ratio of roughly 1/2 for ancestral genes and 1 for newly acquired genes. Taken together, the following picture emerges: ancestral genes need to be compensated in dosage because their concentration is adapted to being present in two copies. Newly acquired genes in contrast are adapted in concentration to the fact that only one X chromosome is active in either males or females and need not be dosage compensated. They however exhibit the full impact of X inactivation in females, whereas this impact may be partly attenuated for ancestral genes that are also subject to upregulation in females.

How do our observations relate to the hypotheses regarding X upregulation that were put forward in the literature? Broadly, these hypotheses can be classified into the no upregulation, global upregulation and partial upregulation classes. Our observations are in line with the partial upregulation of ancestral genes, but furthermore we also observed indications for a precompensation of acquired genes. Thus, we found evidence for global upregulation during differentiation. Consequently, our results are apparently contradicting not only to the hypothesis of no upregulation but also that upregulation is restricted to a small set of dosage sensitive genes.

However, all studies finding no dosage compensation investigated somatic mammalian tissue only (See Tab. 2.1). Thus the results presented above are complementary to these studies since we investigated mouse embryonic stem cells and early differentiation. It follows that in principle our observations and the studies claiming no compensation can be reconciled, by assuming a transient dosage compensation. At least for the ancestral genes, the mechanism responsible for upregulation could in principle be reversible. Also, it is known that X:A ratios vary from tissue to tissue, with brain typically having comparably high X:A ratios (Nguyen and Disteche, 2006; Deng et al., 2011). This leaves the possibility of a fine tuning of the X dosage during differentiation into different tissues.

Although there is no direct contradiction between our results and that of studies claiming no upregulation of the X chromosome, some of the issues raised in these studies also apply to our study. Among these issues, we identified three possibly impacting some of our results: implicit filtering of genes by the platform used for assaying the transcriptome, explicit filtering of lowly expressed genes and the compression of expression differences by microarrays (Tab. 2.2).

By comparing genes detected by the Affymetrix exon arrays and the RNA-seq data set, we showed that the impact of implicit filtering can be limited by restricting the analysis to protein coding genes. However, both for the exon array platform and for the RNA-seq-based data we cannot exclude that we overestimate X:A ratios since the detection ratio is smaller for the X chromosome than for the autosomes. To estimate the X:A ratio in undifferentiated cells, we used an explicit expression threshold. To check whether this alters the X:A ratio, we tested the impact of different expression thresholds. This analysis showed that the assertions for undifferentiated stem cells are not sensitive to the threshold.

Finally, the microarray data we used might underestimate differences in expression between sets of genes and changes of genes through time (Perkins et al., 2014). Consequently, acquired genes may also be upregulated but we were unable to detect this. Also, the precompensation of acquired genes may be an artifact of the compression of differences. However, the evidence from the RNA-seq data points

Table 2.2: Sensitivity of assertions regarding X dosage to gene filtering and platform properties. The linear ratio between the X chromosome and the autosomes is denoted by X:A and XX:AA in males and females, respectively. Assertions applying to undifferentiated mESC are indicated by the day of differentiation (d0). The issues that may influence the assertions are (A) Implicit filtering of genes by the properties of the platform and the normalization procedures used, (B) Explicit filtering of lowly expressed genes, (C) The reduced resolution of microarrays.

observation	cell line	sensitive to
X:AA <1 at d0	XO,XY	A,B
XX:AA ≥ 1 at d0	XX	A,B
$(X{:}AA)_{\text{ancestral}}$ increasing	XO,XY	
$(X{:}AA)_{\text{acquired}}$ constant	XO,XY	C
$(X/XX{:}AA)_{\text{acquired}} > (X/XX{:}AA)_{\text{ancestral}}$ at d0	XO,XY,XX	
$(X{:}AA)_{\text{acquired}} \approx 1$	XO,XY	A,B,C

into the same direction for both observations, showing no consistent upregulation and an expression compatible with X:A≈ 1 for acquired genes.

2.2 Progress of differentiation is timed to compensation of X dosage

Parts of this sub-chapter have been published as "The Two Active X Chromosomes in Female ESCs Block Exit from the Pluripotent State by Modulating the ESC Signaling Network" (Schulz et al., 2014). The gene expression data for differentiating XO, XY and XX mESC has been generated by Edda G. Schulz. Additionally, the experiments shown in Fig. 2.15, 2.17 and 2.18 have been performed by Edda G. Schulz. Bioplex assays shown in Fig. 2.16 have been performed by Anja Siebert. The analysis of target gene expression (Fig. 2.15c) has been performed by Nils Blüthgen. Figures 2.15-2.18 have been prepared by Edda G. Schulz, who was also involved in optimizing the layout of figures 2.11-2.14. Figures 2.11-2.18 are reprinted with slight alterations from Schulz et al. (2014).

2.2.1 Introduction

In the preceding chapter we investigated how the X chromosome dosage changes during early differentiation of mESCs. The major process impacting X chromosome dosage in female cells is X inactivation. It is known that this process is controlled by the core pluripotency factors Oct4, Sox2 and Nanog, which together repress Xist (Navarro et al., 2008). The general consensus is that an initial drop in pluripotency factors leads to a derepression of Xist which in turn triggers X inactivation. This mechanism ensures that X inactivation is timed to the onset of differentiation.

During embryogenesis of female mice, the paternal X chromosome that had been inactivated before is reactivated in the inner cell mass of the blastocyst. As development of the embryo proceeds, one of the two X chromosomes is randomly inactivated. During the time window between reactivation of the paternal X chromosome and random X inactivation, there exists a dosage difference of X-linked genes between female and male embryos. Whether, and if, how this dosage difference impacts the pluripotency-maintaining network and with it the process of differentiation is unclear.

There is some indirect and direct support for the existence of a feedback of X dosage on differentiation. Indirect support comes from the fact that the persistence of double X dosage leads to death of the embryo around E10 (Takagi and Abe, 1990). Thus, a feedback mediating a developmental checkpoint would be beneficial in adding robustness to the process of differentiation. Second, there is evidence for accelerated development of post-implantation embryos with a single X chromosome (Burgoyne et al., 1995; Thornhill and Burgoyne, 1993).

The dosage difference of X-linked genes exists during a period in which critical changes take place in the embryo (see Fig 1.1, page 2). At the same time, pluripotency-associated factors are downregulated, developmental genes are upregulated and *de novo* DNA methylation as well as other epigenetic changes take place. The mechanisms mediating these changes are coupled to each other, ensuring an orderly progression of differentiation through various stages. Despite this coupling, the gene regulatory network of pluripotency-associated factors occupies a central place and is considered to be central for the decision to exit the pluripotent state.

Thus, if the double X dosage in female cells affects differentiation, this effect has to be visible at the level of the pluripotency-associated factors. But furthermore, due to the tight coupling described above differences should be observed at many levels, including widespread differences in gene expression and epigenetic marks. Studying the coupling of X inactivation and the pluripotency-maintaining network thus promises to yield insights not only about the interplay of these two processes but also about other regulatory mechanisms that shape gene expression during differentiation as well as the interdependence of these regulatory mechanisms among each other.

We used the transcriptome time series for the XX, XO and XY cell lines, that was already presented in the preceding chapter, to investigate the impact of X chromosome dosage on differentiation. From global expression patterns in the three cell lines we deduced the existence of a differentiation delay in female cells. By correlating the observed expression patterns with genome-wide data sets of DNA methylation and by comparing the activity of different signaling pathways, we were able to establish a possible regulatory route from the X dosage back to the core pluripotency factors. The existence of this route is confirmed in two ways. First, by signaling perturbation experiments and second, by inducing X inactivation in an Xist-inducible cell line. We could show that speeding up X inactivation indeed alleviates the delay in differentiation found in female cells.

2.2.2 Female mESC are delayed in differentiation

We first aimed to find groups of genes that are differentially expressed between the cell lines during the exit from the pluripotent state. Since we are interested in a dynamic problem, the usual methods for finding differentially expressed genes should be discarded in favor of methods more suited to the analysis of time series. We proceeded in two steps: first, we identified the subset of genes regulated over

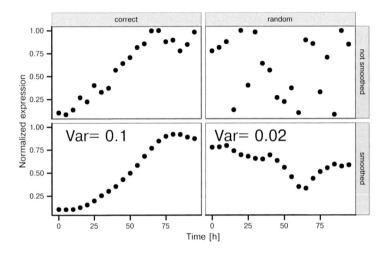

Figure 2.10: Schematic figure illustrating the test for determining regulated genes. Upper panel (not smoothed): The original time order (left panel) of the measured expression values is shuffled to obtain the random expression kinetic (right panel). Lower panel: Both expression kinetics are smoothed with a time window of 24 h. Subsequently, the variance of both kinetics is computed. This is repeated for many shufflings and the variance of the original kinetic is compared to the distribution of variances for the shuffled kinetics.

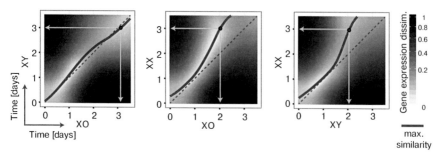

Figure 2.11: The XX cell line is delayed in expression relative to the XO/XY cell lines. Genes are interpolated to a resolution of one hour and for each pair of time points, the Euclidean distance in the space spanned by the expression of all regulated genes is computed. The heat maps indicate the Euclidean distance with maximum distance normalized to 1, while the red line connects points of minimal distance between corresponding samples (See Materials and Methods, Sec. 6.2, page 134). The dashed diagonal corresponds to sample pairs taken at equal time points. The figure is based on the transcriptome time series schematically depicted in Fig. 2.1, page 19.

time using a new non-parametric test. Second, we quantified global expression differences among all regulated genes to identify similarities and points of divergence in expression for the different cell lines.

To identify the regulated genes, we defined a new test statistic. We are faced with the problem to test for differential regulation over time when no replicate measurements are present. The idea of our test is to look at trends in expression over a couple of measurements and compare these trends to the inherent variability of the time series (Fig. 2.10). We proceed as follows: As the test statistic, we define the standard deviation of the smoothed time series with a smoothing window of 24 h. For the null hypothesis we assume that the gene expression does not change on the time scale of 24 h. The distribution of the test statistic under this null hypothesis is calculated by computing the variance on the smoothed, randomly shuffled time series. We term a gene as regulated if the test statistic exceeds all values of the null distribution obtained for 10000 reshufflings, corresponding to a p-value $< 10^{-4}$ (For further details, see Materials and Methods, Sec. 6.2, page 130). In the following analysis, we included all genes that are regulated in either the XX or the XO time series.

We next aimed to quantify global differences in expression between the cell lines. In order to do so, we calculated the Euclidean distance between different samples

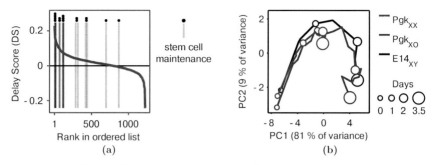

Figure 2.12: Stem cell maintaining genes are delayed in expression in XX cells. (a) GSEA analysis of genes ranked by delay score. Genes with the GOBP annotation stem cell maintaining are indicated by gray lines with dots. (b) Principal component analysis on the gene expression of all regulated genes annotated as stem cell maintaining. The principal components were calculated from the samples of all three cell lines.

in the space spanned by the expression of all regulated genes. For the comparison XO vs. XY, expression is most similar at matching time points. For XX vs. XO or XY, late time points in XX are most similar in expression to intermediate time points in XO or XY (Fig. 2.11). In summary, overall expression changes happen at a similar speed in XO and XY cells whereas XX cells are delayed with respect to the former two cell lines.

To investigate whether the observed delay in XX cells can be connected to the progress in differentiation, we analyzed the delay at the level of individual genes. We defined a delay score for each gene that compares the rate of change in the XO and XX time series, such that a positive delay score implies a larger rate of change in XO cells than in XX cells (For details, see Materials and Methods, Sec. 6.2, page 131). We used the gene set enrichment analysis (GSEA) (Subramanian et al., 2005) to find sets of genes sharing the same biological process GO annotation (GOBP) that are enriched for high or low delay scores (Fig. 2.12a).

The only GOBP terms significantly enriched (at an FDR < 0.05) among genes with high delay were "stem cell maintenance" and "stem cell development", supporting the connection of the delay with the progress in differentiation. To compare this progress in the three cell lines, we performed a principal component analysis (PCA) on the subset of genes annotated with the term "stem cell maintenance" (Fig. 2.12b). The overlapping trajectories show that all three cell lines progress through similar expression states during differentiation. However, though the expression states of the XX cell line are similar to those of the XO and XY cell lines,

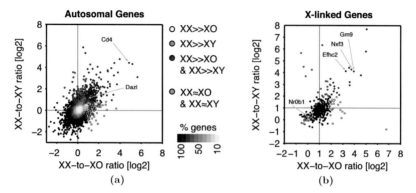

Figure 2.13: A small minority of autosomal genes are affected by the X dosage and some X-linked genes show higher than expected dosage differences between the cell lines. (a) For all autosomal genes, the ratio between XX and XY cells is plotted against the ratio between XX and XO cells. Genes that significantly exceed the expected ratio of 1 (0 in log2 units) are colored in yellow, red and orange. (b) For X-linked genes, the expected ratio is assumed to be 2 (1 in log2 units) and genes for which both ratios are not significantly deviating from a ratio of 1 (0 in log2 units) are colored blue.

progression through these states slows down and comes almost to a halt at day two.

The observed expression trajectories are consistent with the interpretation that while differentiation corresponds to progression through well defined expression states, the speed of progression can be modified by other processes. To prove that this speed is modulated by the state of X inactivation, we have to analyze if and by which mechanisms X dosage feeds back into the core pluripotency network.

2.2.3 Methylation is linked to differentiation delay

We next analyzed the connection between X dosage and DNA methylation. We did so for three reasons: first, cells with two active X chromosomes were found to have globally reduced DNA methylation (Zvetkova et al., 2005), which is likely due to decreased expression of *de novo* methyltransferases (Habibi et al., 2013; Ooi et al., 2010). Second, DNA methylation is an important regulator of expression in undifferentiated mESCs, modulating the expression of hundreds of genes (Karimi et al., 2011). Third, pluripotency is associated with global DNA hypo-methylation (Leitch et al., 2013). Because of this association, we speculated that a lack of *de novo* methylation could be responsible for expression differences accumulating over the course of differentiation.

Since the presence of two active X chromosomes reduces DNA methylation, differences in expression caused by hypo-methylation should be detectable until X inactivation has been concluded. Thus, we should observe differences in expression between XX and XO/XY cells associated with DNA methylation both in undifferentiated and in differentiating mESCs. We first asked if differential expression between XX and XO/XY cells in the undifferentiated state is associated with DNA methylation. In a second step we analyzed the connection between delayed differentiation in XX cells and *de novo* methylation. We investigated both questions by looking for statistical associations between published data on methylation sensitive genes and genes we found to be differentially expressed between XX and XO or XY cells.

To start this investigation, we need to take stock of genes that are differentially expressed between XX and XO or XY cells in the undifferentiated state. Only genes that are sufficiently highly expressed so that expression differences can be resolved by the employed micoarrays were analyzed for differential expression (For details, see Materials and Methods, Sec. 6.2, page 131). We found that the majority of autosomal genes showed fold changes around 0 for XX-to-XY and XX-to-XO, while a small minority (322 out of 9338, 3.4 %) were expressed at significantly higher levels in XX cells than in XO or XY cells (Fig. 2.13a). For X-linked genes, we have to take into account the overall expression ratio of 2 (1 in log2 units) between XX and XO or XY cells (Compare section 2.1.2). Because of this, we term X-linked genes differentially expressed that deviate significantly from an XX to XO or XY fold change of 1 in log2 units. Only a small number of X-linked genes were expressed at significantly lower levels than expected in XX compared to XO/XY cells (19 out of 355, 5.4 %). The same number of genes was expressed at significantly higher levels (19 out of 355, 5.4 %), see Tab. 2.3 and Fig. 2.13b. We call the autosomal and X-linked genes upregulated in XX cells according to the above critera with respect to both XO and XY cells X-dosage-sensitive genes.

Having established the genes that are differentially expressed between the three cell lines in the undifferentiated state, we tested whether differential expression is associated with methylation. More precisely, we asked whether genes that are expressed higher in XX cells than in XO or XY cells are sensitive to methylation at their promoters. In order to test this, we obtained transcriptome data from three previously published studies of DNA methylation deficient triple knockout (TKO) cells (For details, see Materials and Methods, Sec. 6.2, page 134). These cells lack

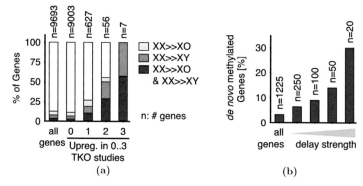

(a) (b)

Figure 2.14: (a) Genes that are derepressed upon loss of DNA methylation are enriched for X-dosage-sensitive genes. Genes were grouped into four groups depending on the number of TKO studies they were found to be upregulated in. In each group, the fraction of X-dosage-sensitive genes (red) and of genes with higher expression in XX than in either XO (yellow) or XY (orange) is indicated. The number of genes in each group is indicated above the respective bar. (b) Strongly delayed genes are enriched for *de novo* methylated genes. Genes were sorted by delay score and the fraction of *de novo* methylated genes was computed for different numbers of top delayed genes. The bars indicate the fraction of *de novo* methylated genes for the indicated number of top delayed genes.

Table 2.3: Number of expressed genes that are differentially expressed in XX cells vs. XO or XY cells grouped according to their relative expression in the different cell lines. Note that some categories are not mutually exclusive.

X-linked genes		Autosomal genes	
Criterion	n	Criterion	n
$XX \gg 2 \times XO$	44	$XX \gg XO$	767
$XX \gg 2 \times XY$	38	$XX \gg XY$	706
$XX \gg 2 \times XO$ & $XX \gg 2 \times XY$	19	$XX \gg XO$ & $XX \gg XY$	322
$XX \ll 2 \times XO$	35	$XX \ll XO$	193
$XX \ll 2 \times XY$	50	$XX \ll XY$	414
$XX \ll 2 \times XO$ & $XX \ll 2 \times XY$	19	$XX \ll XO$ & $XX \ll XY$	78
All expressed genes	355	All expressed genes	9338

DNA methyltransferases ($Dnmt1^{-/-}Dnmt3a^{-/-}Dnmt3b^{-/-}$), so that genes targeted by DNA methylation are upregulated with respect to the wild type. High-confidence targets of DNA methylation identified by upregulation in at least two studies were strongly enriched in X-dosage-sensitive genes (Fisher's exact test, $p < 10^{-15}$) (Fig. 2.14a). We concluded from this strong association that DNA hypomethylation constitutes a possible mechanism for derepression of genes in female mESCs.

As indicated above, we additionally speculated that the differences in expression accumulating over the course of differentiation may be caused by less *de novo* methylation in females than in males. According to this hypothesis, we should find an enrichment of *de novo* methylated genes among genes with strong delay. To test this hypothesis, we obtained genome-wide DNA methylation profiles comparing mESCs with neural precursor cells (Stadler et al., 2011) (For details, see Materials and Methods, Sec. 6.2, page 133). We found that the fraction of *de novo* methylated genes increases from 3 % for all regulated genes to 30 % for the 20 most delayed genes (GSEA, $p < 10^{-3}$) (Fig. 2.14b).

Taken together, we identified DNA hypomethylation as a likely candidate for causing large scale gene expression differences in response to two active X chromosomes. A lack of DNA methylation is associated both with expression differences between XX and XO/XY cells in the undifferentiated state and with the delayed kinetics of genes over the course of differentiation.

2.2.4 X dosage and methylation are connected via MAPK signaling

So far, we used statistical analyses to find patterns in our transcriptome data and correlate these patterns to other high throughput data. This approach allowed us to detect the global delay in differentiation in female cells and to connect this delay to hypomethylation in these cells. The goal of our investigation, however, was to answer the question if and by what mechanism the X dosage feeds back into the core pluripotency network, modulating the speed of differentiation.

Showing that this feedback exists involves two critical steps. First, we need to find the intermediate regulatory processes that constitute the mechanistic link between X dosage and methylation. Second, it remains to be shown that the forcing of X inactivation speeds up differentiation in female cells. Both of these points can only be investigated experimentally and are thus not in the focus of this work. For this reason they will be presented in a condensed manner.

In the following, we will argue that MAPK signaling connects X dosage and DNA methylation, yielding the link we are searching for. The connection of MAPK signaling with pluripotency has been uncovered in the process of improving culture conditions for embryonic stem cells. Traditionally, cells were cultivated in serum-based medium that led to a heterogeneous mixture of undifferentiated and primed stem cells (Graf and Stadtfeld, 2008).

The refinement of culture conditions eventually led to the discovery of the 2i medium that consistently keeps cells in a developmental ground state by inhibiting both the Wnt pathway component Gsk3 and MAPK pathway component Mek (Nichols and Smith, 2009). The crucial role of MAPK signaling in pluripotency thus makes it a candidate for mediating differences in the progress of differentiation between XO/XY and XX cells. To investigate this role, it is necessary to employ serum-based culture conditions in order not to directly interfere with MAPK signaling.

We first noticed that in serum, expression of pluripotency-related factors is consistently higher in XX mESCs than in XO or XY mESCs, pointing to a more undifferentiated state in XX cells (Fig. 2.15a). These differences are eliminated in 2i medium (Fig. 2.15b), supporting our speculation that X dosage might influence the signaling pathways inhibited by this medium and in turn the pluripotency factors. Following this idea, we checked whether target genes of several signaling pathways are influenced by the X dosage. In serum, we found that indeed, targets of Mek and Gsk3 had a lower expression in XX cells than in XY cells, while targets of Akt were upregulated. In 2i medium, differences in expression between the cell lines were much less pronounced (Fig. 2.15c).

To confirm that the observed expression differences are indeed due to signaling activity, we also quantified relative phosphorylation levels of signaling pathway components (Fig. 2.15d). We found phosphorylation levels of Mek and Erk2 as

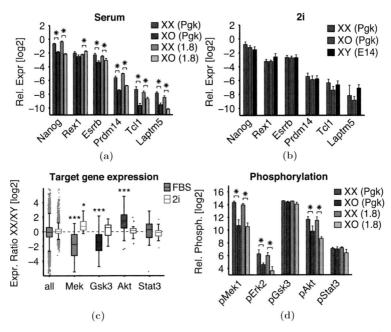

Figure 2.15: X dosage alters activity of several signaling pathways. The relative expression of pluripotency-associated genes was quantified by qPCR in mESC of the indicated cell line grown in serum-based medium (a) and in 2i medium (b). (c) The expression ratio of targets of the indicated signaling pathways in Pgk XX versus E14 XY mESCs grown in 2i medium (open boxes) or serum (filled boxes) is shown. Boxes indicate the interquartile range, horizontal bars indicate the median. Targets of signaling pathways were identified with the help of published transcriptome data obtained after inhibiting the respective pathway. Expression ratios for 2i were computed using the data from undifferentiated XO, XY and XX cells in our transcriptome time series and published transcriptome data for expression in serum (Marks et al., 2012). (d) Phosphorylation levels of Mek, Erk2, Gsk3, Akt and Stat3 were quantified in XX and XO mESC using Luminex technology.

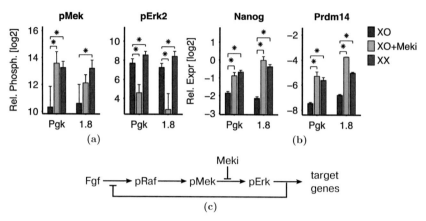

(a)   (b)

(c)

Figure 2.16: (a) Phospho-Mek and Phospho-Erk2 levels in XO mESCs (blue), XO mESCs incubated for two days with a Mek inhibitor (light blue) and XY mESCs (red). (b) Mek inhibition restores Nanog and Prdm14 levels in XO mESCs to concentrations seen in female (XX) mESCs. Relative expression quantified by qPCR in in XO mESCs (blue), XO mESCs incubated for two days with a Mek inhibitor (light blue) and XY mESCs (red). (c) Scheme of the MAPK pathway with the action of the Mek inhibitor (Meki) and the negative feedback of Erk on upstream components indicated. The species pRaf, pMek and pErk indicate the active, phosphorylated form of the proteins.

well as those of Akt to be elevated in XX cells relative to XY cells. While we expected phospho-Akt levels to be increased in XX cells, elevated phosphorylation of Mek and Erk2 seems to contradict reduced gene expression of targets of the MAPK pathway. This contraintuitive result can be explained with the presence of a negative feedback from Erk to components upstream of Mek (Fig. 2.16c). Inhibiting Mek catalytic action reduces Erk activity and relieves the negative feedback of Erk on upstream components, leading to an increase in Mek activity (Fritsche-Guenther et al., 2011; Sturm et al., 2010).

However, in our case this would mean that Erk activity also has to be reduced in XX cells compared with XO or XY cells. To analyze the precise nature of the feedback in mESC, we treated XO cells with a Mek inhibitor for two days and compared the treatment with activity of Mek and Erk in XX cells. We observed that phospho-Mek levels could be restored to levels found in XX cells by inhibitor treatment. In contrast, phospho-Erk2 levels were reduced by the inhibitor treatment in XO cells but are high in untreated XO cells and XX cells (Fig. 2.16a). Together, our observations show that two active X chromosomes inhibit MAPK

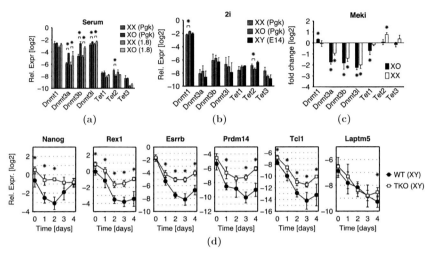

Figure 2.17: Methyltransferase expression is modulated by the MAPK pathway. Relative expression of DNA methylation modifying enzymes was quantified by qPCR in mESC of the indicated cell line grown in serum-based medium (a) and in 2i medium (b). (c) The fold change of expression of DNA methylation modifying enzymes was measured after two days of treating Pgk XX and XO cells with a Mek inhibitor. (d) The expression of selected genes was compared between the Dnmt TKO cell line and the parental cell line that is wild type for the DNA methyltransferases. The mean and standard deviation of three to six experiments are indicated by points and error bars, respectively. Asterisks indicate significant differences at the significance level of $p < 0.05$.

signaling but that the inhibitory effect has to occur more downstream than the action of the Mek inhibitor, leaving Erk activity on a higher level.

Finally, we tested the effect of Mek inhibition on Nanog and Prdm14. According to our hypothesis, inhibiting the MAPK pathway should stabilize the pluripotent state. Indeed, we found that Mek inhibition raises Nanog and Prdm14 expression in XO cells to levels found in XX cells or higher (Fig. 2.16b).

After having established that two active X chromosomes inhibit the MAPK signaling pathway, we tried to integrate the observations connected with MAPK signaling with the observations connected to DNA methylation. As we showed above using statistical analyses, genes that exhibit a delay in up- or downregulation in XX mESCs are often targeted by *de novo* methylation. We also showed that genes expressed higher in undifferentiated XX cells than in undifferentiated XY or XO cells are strongly enriched for methylation-targeted genes. The presence of two active X chromosomes is thus connected with reduced methylation in undifferentiated cells and possibly also with an impaired gain in methylation over differentiation. As we showed above, two active X chromosomes also alter MAPK signaling. Based on these two observations we asked whether there is a causal link between MAPK signaling and DNA methylation which would yield a causal link between X chromosome dosage and DNA methylation.

We first observed that the DNA methyltransferases Dnmt3a, Dnmt3b and Dnmt3l exhibit elevated expression in XO cells relative to XX cells in serum (Fig. 2.17a), whereas we found no significant expression differences in 2i medium (Fig. 2.17b). Since the differences vanish in 2i medium, we hypothesized that DNA methyltransferases are also regulated by the MAPK pathway. Consistently, we found Dnmt3a, Dnmt3b and Dnmt3l expression to be reduced in XO and XX mESCs after two days of incubating with a Mek inhibitor (Fig. 2.17c). This shows that repressed MAPK signaling activity causes reduced activity of DNA methyltransferases, which leads to DNA hypomethylation.

The impact of DNA hypomethylation on selected pluripotency factors was determined by comparing the expression in Dnmt TKO cells to their parental wild type cell line (Fig. 2.17d). This comparison showed that central pluripotency factors such as Nanog, Esrrb and Prdm14 are significantly less downregulated during differentiation in the cell line lacking the DNA methylation enzymes. Thus, the reduced expression of DNA methyltransferases caused by inhibited MAPK signaling in cells with two active X chromosomes leads to impaired downregulation of pluripotency factors.

We have thus established that two active X chromosomes stabilize the pluripotent state via two routes. First, they inhibit MAPK signaling which prevents pluripotency factors from being downregulated. Second, inhibted MAPK signaling in turn reduces the activity of DNA methyltransferases relative to XO or XY cells, leading to DNA hypomethylation which prevents repression of pluripotency factors.

As a direct proof that X dosage modulates the progress of differentiation, we performed a final experiment where we induced X inactivation. We used the TX1072 cell line, in which Xist can be induced from the endogeneous locus by doxycycline (Schulz et al., 2014). First, we ensured that the construct works as expected by comparing the number of Xist positive cells in cells where doxycycline was added one day prior to the start of differentiation to cells differentiated in the absence of doxycycline (Fig. 2.18a).

Next we checked whether the faster upregulation of Xist we observed in induced cells leads to a faster downregulation of pluripotency maintaining factors. We

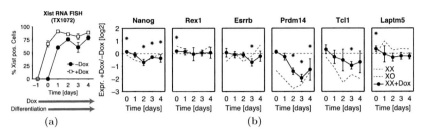

Figure 2.18: (a) Xist RNA fish assay in cells with an inducible promoter at the transcriptional start site of the endogenous Xist gene differentiated in the presence (+Dox) or absence (-Dox) of doxycycline. The line and the error bars indicate the mean and standard deviation of three independent experiments.(b) Expression levels of selected genes which were quantified using qPCR in the presence (black) and absence (red) of doxycycline and in a TX1072 XO line (blue). Expression was normalized to untreated XX cells (red). Error bars indicate the mean and standard deviation of three to five independent experiments, asterisks indicate significant differences at the significance level of $p < 0.05$.

compared XX cells that were not induced, XX cells incubated with doxycycline and XO cells. Consistent with our transcriptome data, downregulation of pluripotency associated factors is stronger in XO cells than in XX cells. However, XX cells with doxycycline induced Xist upregulation exhibit a downregulation that is nearly as strong as that found in XO cells (Fig. 2.18b). This finding proves that the exit from the pluripotent state can indeed be accelerated by increasing the speed of X inactivation.

2.2.5 Discussion and outlook

The process of X inactivation affects the expression of hundreds of X-linked genes. Because of this widespread change in gene expression, X inactivation has to be tightly controlled by the pluripotency-maintaining network. This need is evidenced by the control of Oct4, Sox2 and Nanog over Xist itself and factors modulating Xist expression. Combining statistical analysis of transcriptome data of differentiating mESC and experimental evidence, we showed that the X dosage feeds back into the network of pluripotency-associated factors. Incomplete X inactivation slows down differentiation, making proper X shutdown a developmental checkpoint.

Our analysis of the transcriptome of differentiating mESC showed that female cells are delayed in differentiation with respect to male (XO or XY) cells. Because we found strongly delayed genes to be enriched for genes that are *de novo* methylated during differentiation, we hypothesized that the presence of an addi-

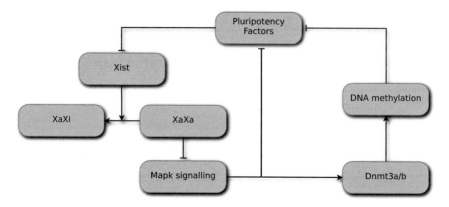

Figure 2.19: Model of the interaction between pluripotency factors and X inactivation.

tional X chromosome represses methylation and thus progress in differentiation. Experimental evidence established that two active X chromosomes repress MAPK signaling which in turn activates DNA methyltransferases. The feedback of X dosage on the pluripotency-associated factors via MAPK signaling and methylation was confirmed by an experiment where forced induction of Xist increased the speed of differentiation in female mESC.

We can sum up the above findings in an extended model of the interplay of differentiation and X inactivation (Fig. 2.19). In the undifferentiated state, pluripotency factors are highly expressed, repressing Xist so that the two active X chromosomes inhibit MAPK signaling. This in turn stabilizes the pluripotent state by keeping DNA *de novo* methylation repressed. Once differentiation is triggered, decreasing pluripotency factors lift the repression of Xist, leading to X inactivation and in turn derepression of the MAPK pathway. The MAPK pathway acts both directly by repressing pluripotency factors and indirectly by upregulating DNA methyltransferases, causing further repression of the pluripotency factors by methylation of their promoters. As the downregulation of the pluripotency factors proceeds, cells exit the pluripotent state.

The most pressing open question remaining is which gene(s) on the X chromosome transmit the information about the X dosage to the pluripotency-maintaining network. Interesting candidates are the X-linked pluripotency factor Nr0b1 or the Erk phosphatase Dusp9, which also lies on the X chromosome. However, so far none of these genes could be shown to play the role of the crucial link between X dosage and differentiation (Schulz et al., 2014).

The search for the gene mediating this link could be aided by a bioinformatical analysis that would predict candidate genes for an experimental follow-up analysis. A candidate gene would have to be X-linked, sensitive to X chromosome dosage in undifferentiated stem cells and responsive to X inactivation. The search for the missing link thus calls for an approach that continues to alternate between statistical analyses of high-throughput data, small scale assays and perturbation experiments.

Our analysis shows that differentiation is not an autonomous process with a clear hierarchy where pluripotency-associated transcription factors occupy the top position. In contrast, the pluripotency-maintaining signaling and gene regulatory network is also involved into at least one (likely one of many) feedback loop. This feedback indicates the complexity of gene regulation during differentiation. In the next chapter, we will study a mechanism that mediates coregulation of neighboring genes, giving rise to functional modules of genes. Such functional modules can limit the complexity of gene regulation mentioned above. In particular, we will show that the spatial architecture of the genomic neighborhood of Xist shapes the response to trans-acting initiation of X inactivation by pluripotency-associated factors.

3 A mechanism for coordinated expression of adjacent genes

Parts of this chapter have been published as "Spatial partitioning of the regulatory landscape of the X-inactivation centre" (Nora et al., 2012). The 5C-data for the analysis of the X inactivation center shown in Fig. 3.1 was generated and analyzed by Elphege P. Nora, Bryan R. Lajoie and Nynke L. van Berkum. TAD borders (Tab. 6.4) were annotated by Elphege P. Nora. Experimental evidence discussed in section 3.3 (RNA fish data for Fig. 3.5 and transgene annotation, Fig. 3.4) was generated by Elphege P. Nora. The gene expression data for differentiating XO, XY and XX mESC was generated by Edda G. Schulz. Figures 3.1, 3.2, 3.4 and 3.5 are reprinted from (Nora et al., 2012), with additional information added in Fig. 3.4 by the author.

3.1 Introduction

In the preceding chapter we studied the chromosome-wide effects of X inactivation on X-linked genes. Although certain genes may escape X inactivation, it is a process affecting large sets of X-linked genes in a similar manner. It is thus an example of a regulatory process that operates relatively indiscriminately and globally. On the other end of the spectrum we find promoter-specific regulation of single genes, that can in principle lead to unique expression patterns for a gene. In terms of length scales, in this case the regulatory region (the promoter) is directly neighboring the coding sequence of the gene. In the following we will study a regulatory mechanism operating at an intermediate level in terms of specificity and range that gives rise to coordinated expression of neighboring genes.

It is well established that gene order in prokaryotes is not random (Price et al., 2005). Functionally related genes are organized in operons that allow for a co-ordinated expression of these genes. It is much less clear how widespread such a clustering of functionally related, coexpressed genes is in eukaryotes. However, in eukaryotes so called "position effects" - the influence of the genomic location of a gene on its expression - were observed. These effects were analyzed using transgenes, whose expression depends on the location of the integration site (Milot et al., 1996). As we would expect from this observation, adjacent genes tend to be coexpressed. However, the functional role of these neighbor-to-neighbor effects are not completely understood.

It is tempting to think that functional modules, in which genes share function and expression context, also have to exist in eukaryotes. These modules might decrease concentration mismatches of interacting genes and facilitate evolution by providing self-contained units. There is limited evidence for gene clusters forming functional modules in eukaryotes, for instance clusters of genes involved in the same pathway (Lee and Sonnhammer, 2003). Well documented functional modules such as Hox and β-globin genes, however, are rare (Hurst et al., 2004). Thus, mechanisms that can coordinate gene expression have to be studied further to find out whether eukaryotic expression modules are a rare exception or a commonly encountered feature.

What are the mechanisms for coregulation of neighboring genes? Genes may share a common bi-directional promoter (Li et al., 2006) or co-expression may be induced by the spread of histone modifications (Hurst et al., 2004). Recently, investigations into the 3d organization of the genome have established the regulatory role of chromatin contacts. At a range of typically 10 to 100 kb, enhancer-promoter interactions lead to local chromatin loops that may mediate gene co-expression by reducing the distance between promoters located in the loop. At much larger distances, the eukaryotic genome tends to segregate into two compartments, corresponding to open, transcriptionally active, and closed, transcriptionally inactive, chromatin (Lieberman-Aiden et al., 2009). Transition of chromatin regions between the two compartments may lead to coordinated up- or downregulation of the genes contained in that region.

Here, we study the spatial organization of the X-inactivation center (Xic) in order to investigate how chromatin contacts influence gene expression during differentiation. We employ data obtained using the chromosome conformation capture carbon-copy (5C) technique that allows to interrogate chromatin contacts at a high resolution (Dostie et al., 2006).

This data shows that the region around the Xic is partitioned into discrete domains of preferential chromatin contact that have a size between 200 kb and 1 Mb. We show that these topologically associated domains (TADs) align with clusters of coregulated genes. This alignment observed around the Xic lead us to speculate that TADs may be a mechanism driving co-expression of functionally related

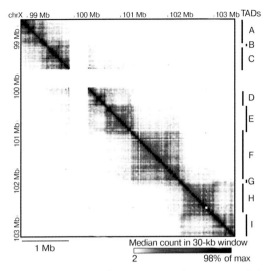

Figure 3.1: Chromosomal contact frequencies in the vicinity of the X inactivation center form discrete blocks. Contact frequencies were determined by 5C in E14 XY mESC. The heat map shows median counts in 30 kb windows shifted by a 6 kb step size.

genes, giving rise to functional modules. Using published chromosome conformation data from a Hi-C experiment, we investigated this hypothesis genome-wide. We observed that gene pairs in TADs show increased co-expression during differentiation compared to other gene pairs. In addition to expression data, we also used protein-protein interaction data to functionally characterize genes in TADs. Using this data, we observed that proteins encoded by distant genes are more likely to interact when they are in the same TAD than when they are not in the same TAD.

3.2 Topological domains align with co-expression modules at the X inactivation center

In order to investigate the impact of chromatin contacts on expression during X inactivation, we used a detailed contact map generated using 5C in mouse embryonic stem cells (mESC) across a 4.5 Mb region containing Xist. This contact map exhibits a discrete block structure with blocks of length 0.2-1 Mb, characterized by preferential long range (> 50 kb) contact among the contained sequences (Fig.

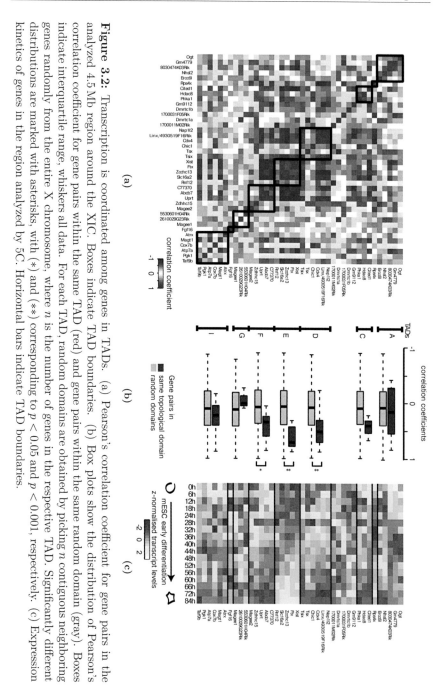

(a)

(b)

(c)

Figure 3.2: Transcription is coordinated among genes in TADs. (a) Pearson's correlation coefficient for gene pairs in the analyzed 4.5 Mb region around the XIC. Boxes indicate TAD boundaries. (b) Box plots show the distribution of Pearson's correlation coefficient for gene pairs within the same TAD (red) and gene pairs within the same random domain (gray). Boxes indicate interquartile range, whiskers all data. For each TAD, random domains are obtained by picking n contiguous neighboring genes randomly from the entire X chromosome, where n is the number of genes in the respective TAD. Significantly different distributions are marked with asterisks, with (*) and (**) corresponding to $p < 0.05$ and $p < 0.001$, respectively. (c) Expression kinetics of genes in the region analyzed by 5C. Horizontal bars indicate TAD boundaries.

3.1). With a size in the sub-megabase range, these blocks are considerably smaller than domains of open or closed chromatin, called type A and B compartments. In the following, these blocks will be called topologically associated domains (TADs). The TAD borders were determined by hand from the normalized 5C data (Nora et al., 2012) (For the coordinates of the TADs, see Materials and Methods, Sec. 6.3, page 135).

To investigate the functional relevance of the sub-megabase chromatin folding structure during X chromosome inactivation, we checked whether expression patterns in female mESC align with the TAD structure. To do this, we calculated the Pearson correlation among all genes that are not pseudogenes in the region analyzed by 5C using the expression data for XX mESC already presented in chapter 2 (For details regarding the determination of gene positions, see Materials and Methods, Sec. 6.3, page 135). The correlation matrix shows blocks of highly correlated genes aligning with the TAD borders (Fig. 3.2a, Fig. 3.2c). This effect is most pronounced for the TAD containing Xist and the two adjacent TADs.

We next checked the statistical significance of this observation in two steps. First, we asked whether this correlation among blocks of adjacent genes could arise by chance. Second, we investigated to which degree the observed correlations can be explained by neighbor-to-neighbor correlations among arbitrary genes that are not necessarily in the same TAD.

To assess the significance of the observed intra-TAD correlations, we set up a simple null model. For each TAD, we constructed 1000 random blocks of adjacent genes on the X chromosome containing an equal number of genes as the TAD. We then compared the distribution of correlation coefficients in each TAD to those of the corresponding random domains using a two-sided Wilcoxon rank-sum test. For three out of seven domains, we find that correlation among gene pairs within a TAD is significantly higher than for gene pairs in random domains (Fig. 3.2b). While this means that the majority of domains does not show a significant increase, two factors have to be taken into account. First, TAD C contains only three genes, which is too small to detect a significant effect though correlation is cleary increased. Second, TAD A and TAD F show some substructure in the correlation matrix (Fig. 3.2a) and also in the chromosomal contact matrix (Fig. 3.1). This points to the possibility of alternative domain definitions, which will be explored in the genome-wide analysis presented below.

It has been previously described that neighboring genes in eukaryotes are correlated in expression (Trinklein et al., 2004; Michalak, 2008; Tsai et al., 2009). This implies that some part of the observed co-expression is due to this general

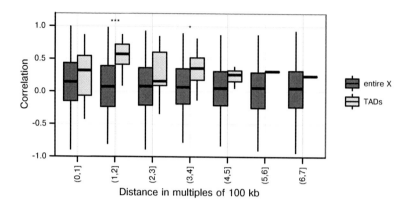

Figure 3.3: Correlation in TADs is higher than average correlation of neighboring genes on the X chromosome. Distances and average correlations have been computed for gene pairs in the same TAD (light gray) and arbitrary gene pairs on the X chromosome (dark gray). Boxes and whiskers indicate the interquartile range and 1.5 times the interquartile range, respectively. Significantly different distributions are marked with asterisks (Wilcoxon's rank-sum test), with ($*$) and ($***$) corresponding to $p < 0.05$ and $p < 0.001$, respectively

effect. While we already compared the extent of co-expression in TADs to that of neighboring genes with the random model, we did not control for the size of these random domains in terms of length of the contained sequence. To include this information, we asked how the correlation of gene pairs in TADs compares to the typical correlation of a gene pair on the X chromosome at a given distance.

We first compared the observed correlation of arbitrary gene pairs on the X to previous observations in the literature. Although differences exist between species, mouse and human were found to exhibit similar co-expression of neighboring genes (Fukuoka et al., 2004; Woo et al., 2010). For humans, the correlation coefficient was found to decrease approximately linearly with logarithmic distance, falling below 0.1 at distances larger than 100 kb (Gherman et al., 2009). Consistent with these previous findings, we found that median expression correlation of neighboring genes on the X quickly decreases to less than 0.1 for distances larger than 100 kb (Fig. 3.3).

Comparing this average distance-dependent correlation for arbitrary X-linked genes with that in TADs, we found that the correlation in TADs is increased for all distances (Fig. 3.3). Consequently, correlations between neighboring genes are too weak to account for the increased correlation in TADs. However, the number of domains in the region we analyzed and the number of genes contained in these domains is too small to verify if this effect is statistically significant at all distances. In general, other layers of regulation will obscure any TAD-related co-expression. We thus have to extend the scope of our analysis to investigate the possibility of TADs constituting a general regulatory principle.

3.3 Both the Tsix and Xist module are essential for proper Xist expression

In the previous section we showed that TADs give rise to blocks of coexpressed genes. Can we consider these blocks as functional modules whose genes share a common role in the X inactivation process? To answer this question, we will discuss experimental evidence. This evidence was not generated in the context of this work but will be presented because it gives causal evidence for the impact of topological domains on expression that can not be obtained with a computational analysis.

The TADs showing the most pronounced co-expression are TADs D,E and F. Interestingly, TAD E is the only domain whose genes all get activated over time (Compare Fig. 3.2). This TAD contains Xist as well as other known positive regulators of Xist: the non-coding RNA gene Jpx, the non-coding RNA gene Ftx and the protein-coding gene Rnf12 (Augui et al., 2011). Furthermore, Xist and Rnf12 mark the 5' and 3' end of this TAD, respectively. Since Rnf12 is a known key activator of Xist (Augui et al., 2011), the entire TAD E is necessary for proper upregulation of Xist upon X inactivation.

Conversely, TAD D harbors elements repressing Xist. It contains the promoter of Tsix, the antisense repressor of Xist as well as the Xite enhancer of Tsix (Augui et al., 2011). We speculated that TAD D contains the elements necessary for maintaining Xist repression by Tsix at earlier stages in differentiation. Together, TADs D and E would then form the minimal region needed for proper Xist expression, mediating the switch from initial Xist repression to subsequent Xist upregulation.

To test if TAD D is the minimal region needed for Tsix expression, we employed two different transgenic mouse lines, Tg53 and Tg80. The transgenes have a common 3' end but whereas Tg53 encompasses the whole TAD D, Tg80 is truncated

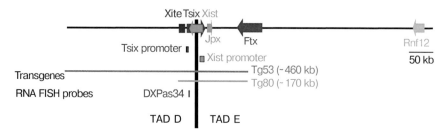

Figure 3.4: Structure of the X inactivation center. Known positive regulators of Xist are Jpx, Ftx and Rnf12. The Xist coding region overlaps with its anti-sense repressor, Tsix, which is activated by the enhancer element Tsx. The vertical line indicates the TAD border between TADs D and E.

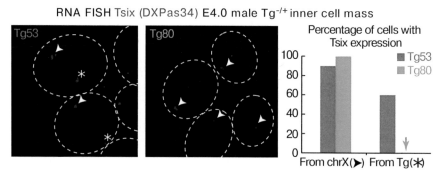

Figure 3.5: The full TAD D, which is adjacent to the Xist-containing TAD E is needed for proper Tsix expression. RNA FISH analysis of modulation of Tsix expression by 5' flanking region. Transgenic Tsix alleles are marked with a star, while endogenous Tsix alleles are marked with an arrowhead. In the inner cell masses of heterozygous transgenic male E4.0 embryos, Tsix expression is only detected from single-copy paternally inherited Tg53 but not Tg80 transgenes.

5' to Xite (Fig. 3.4). We used RNA FISH to detect Tsix expression in the inner cell mass of male E4.0 embryos harboring either the Tg53 or Tg80 transgene. We could only detect Tsix expression from the transgene in Tg53, which covers the entire TAD D (Fig. 3.5). We concluded that TAD D must contain additional elements necessary for upregulation of Tsix and that the entire TAD D is needed for maintaining Xist repression during early stages of differentiation. Thus, TADs D and E organize Xist regulators into functional groups of genes needed for Xist repression and genes needed for Xist activation, respectively.

3.4 Association of co-expression and domains holds genome-wide

So far, our analysis suffers from two limitations. First, the findings were obtained for a region that makes up less than 3 % of the X chromosome. Second, the correlation of gene pairs in TADs scatters considerably, making it hard to estimate the correlation decline with distance. To address both shortcomings we additionally performed a genome-wide analysis of coordinated expression in TADs. The focus of this investigation is the question how strongly the decay of correlation with distance is attenuated for genes in the same TAD.

The co-expression induced by chromatin contacts has already been investigated with low resolution Hi-C data. One study looked at the influence of contact probability between sequences containing promoters on the corresponding genes (Woo et al., 2010). This study found that the correlation of gene pairs on the same chromosome increases with contact probability. However, this analysis fails to control for one dimensional distance along the chromosome. Because of the correlation between genomic distance and 3d distance measured by chromatin contact frequency we have to carefully disentangle the effects of chromatin conformation and 1d distance.

To assess the relevance of topological domains at a genome-wide scale, we obtained chromosome conformation data for mESC that was generated using Hi-C by Bing Ren and coworkers (Dixon et al., 2012) (See Materials and Methods, Sec. 6.3, page 135). This data set contains contact frequencies for sequence pairs at a resolution of 40 kb. In a first step we ignored the fine grained information and made use of the domain annotation supplied along with the data set. In this work, TADs were defined to be delimited by contiguous sequences which first have a bias to bind downstream sequences followed by a bias to bind upstream sequences. The "true" bias was inferred from the underlying states of a Hidden Markov Model,

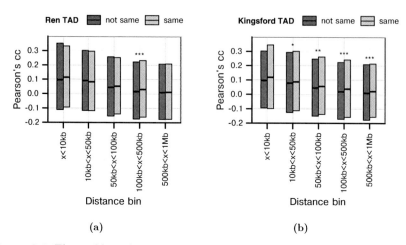

(a)   (b)

Figure 3.6: Elevated long distance gene pair co-expression depends on the domain definition. Gene pairs are classified in two groups, gene pairs in the same TAD and gene pairs not in the same TAD (these include genes that are in TAD free regions or gene pairs from different TADs) and binned according to the distance between the genes making up the pair. Boxes indicate the interquartile range of Pearsons's correlation coefficient and black bars indicate the median correlations. TAD definitions are taken from Dixon et al. (2012) (a) and from Filippova et al. (2014) (b). Significantly different distributions are marked with an asterisk (Wilcoxon's rank-sum test), with (∗), (∗∗), (∗∗∗) corresponding to $p < 0.05$, $p < 0.01$ and $p < 0.001$, respectively.

where each state corresponds to a mixture of Gaussians describing the probability distribution of the binding bias Dixon et al. (2012). The median size of the TADs defined in this way is 880 kb (See Materials and Methods, Sec. 6.3, page 136).

Using these TADs defined by binding direction bias in the Hi-C data, we tested whether gene pairs in the same domain show less correlation decay with distance than gene pairs from different TADs. The correlation of gene pairs on the same chromosome was calculated by taking the mean correlation coefficient for the three time series corresponding to the XO, XX and XY mESC cell lines. We binned gene pairs with distance up to 1 Mb in different distance intervals, less than 10 kb, between 10 kb and 50 kb, between 50 kb and 100 kb, between 100 kb and 500 kb and between 500 kb and 1 Mb. If TADs cause the decrease of co-expression with

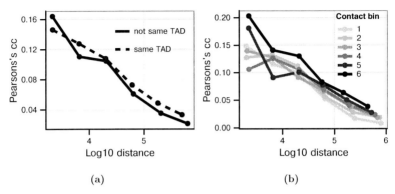

(a)   (b)

Figure 3.7: Attenuated decrease of correlation in TADs is only observed for large distances whereas gene pairs with strong individual contacts have elevated correlation at all distances. Mean distance and mean correlation was computed for gene pairs in equidistant distance intervals of 500 kb length. For the definition of TAD borders, the Kingsford definition has been used.(a) Gene pairs from the same TAD (broken line) are compared with gene pairs from different TADs (solid line). Note that average distance coordinates are shifted to the right for gene pairs from the same TAD. These gene pairs are closer than other gene pairs in the same distance bin. (b) Gene pairs are binned according to contact frequency relative to equally distant pairs from 1 (lowest) to 6 (highest). The contact frequency was obtained from the normalized Hi-C data.

distance to be reduced, we expect the most pronounced effects for the two distance bins corresponding to the largest distance. However, when we tested for increased correlation of gene pairs in TADs we only found a significant increase in the 100 kb to 500 kb distance bin (Fig. 3.6a). Given the larger median size of the TADs of 800 kb we would have expected a significant effect also at larger distances.

The absence of a significant effect for large distances raises the possibility that the partioning algorithm wrongly identifies some large TADs. To assess the effect of the partitioning of the Hi-C data into domains, we compared the above results with those obtained with an alternative algorithm (Filippova et al., 2014) (in the following: Kingsford definition). For this algorithm, an optimal partitioning is defined in dependence of a resolution parameter (See Materials and Methods, Sec. 6.3, page 136). The final partitioning corresponds to the most persistent domains across various resolutions. With a median TAD size of about 160 kb, this partitioning leads to smaller TADs than the ones proposed by Ren and coworkers.

The main biological feature of the alternative domains is that they induce a higher enrichment of CTCF binding sites at the domain boundaries. The protein CTCF is known to be a crucial element in defining TAD boundaries (Ong and Corces, 2014). Consequently, we expected this TAD definition to lead to a more pronounced co-expression effect. Indeed, compared to the TAD definition by Ren et al. (in the following: Ren definition), we found a stronger increase in distance dependent correlation (Fig. 3.6b) for gene pairs in TADs. In particular, we find a significant increase in correlation for the three bins corresponding to distances larger than $50\,\text{kb}$ (Wilcoxon's rank-sum test: $p < 10^{-2}$ for $10\,\text{kb} < x < 50\,\text{kb}$, $p < 10^{-12}$ for $100\,\text{kb} < x < 500\,\text{kb}$ and $p < 10^{-3}$ for $500\,\text{kb} < x < 1\,\text{Mb}$).

The boundaries of TADs are defined based on the matrix that contains the contact frequency for each pair of sequences at a given resolution. Based on the analysis of TADs around the XIC and on the genome-wide data, we proposed that co-expression of genes is induced by spatial proximity of their coding sequences. If this is true, then we should see increased correlation in expression for spatially close genes regardless whether they are in the same TAD or not. To test this, we obtained the contact frequency matrix for each chromosome in mESC from Dixon et al. (2012). We binned gene pairs separately for each genomic distance interval into 6 contact frequency intervals (For details, see Materials and Methods, Sec. 6.3, page 135). With this 1-d-distance-dependent binning, the contact frequency bins measure the relative strength of the contact of a gene pair relative to gene pairs at similar genomic distances.

We next analyzed the average correlation versus genomic distance for each of the contact frequency bins. We observed that higher contact frequencies lead to a consistent shift of the curve to higher correlations, but do not impact the slope of the curve (Fig. 3.7b). This shows that irrespective of the gene block structure defined by TADs, increased contact of genes is associated with increased co-expression. However, by definition regions of strongly increased chromatin contacts will more likely fall into the same TAD and vice versa. The degree of the coupling of these two features can be analyzed by looking at the enrichment of strong contacts in TADs.

We compared the distribution of contact frequencies of gene pairs in TADs with the same distribution for gene pairs not in the same TAD (Fig. 3.8). While gene pairs with strong contacts are highly enriched in TADs at long distances, the distribution remains almost constant up to $10^{4.5} \approx 30\,\text{kb}$. This explains why we see

Table 3.1: Linear models for the correlation coefficient. Gene pairs from the same chromosome up to a distance of 1 Mb were considered. Model (i) allows for a modulation of the decrease of the correlation between gene pairs with distance (dist) by including a dummy variable for TAD membership. The coefficients dist (not TAD) and dist (TAD) indicate the coefficients for the distance variable for gene pairs not in the same TAD and gene pairs in the same TAD, respectively. Model (ii) tests for an additive effect of TAD membership by including the additive dummy variable TAD. Model (iii) corresponds to a linear regression without dummy variable, given by the sum of genomic distance (dist) and contact frequency (bind). The indicated p-values were calculated with a t-test.

Model		Coefficient	Estimate	Std. Error	p-value
(i)	c+TAD*dist	c	2.77×10^{-1}	7.75×10^{-3}	1.01×10^{-278}
		dist (not TAD)	-4.43×10^{-2}	1.38×10^{-3}	1.95×10^{-226}
		dist (TAD)	-4.27×10^{-2}	1.53×10^{-3}	5.58×10^{-172}
(ii)	c+TAD+dist	c	2.72×10^{-1}	8.02×10^{-3}	5.85×10^{-252}
		TAD	8.92×10^{-3}	1.59×10^{-3}	2.07×10^{-8}
		dist	-4.35×10^{-2}	1.42×10^{-3}	5.41×10^{-205}
(iii)	c+bind+dist	c	1.79×10^{-1}	1.26×10^{-2}	4.11×10^{-46}
		bind	2.23×10^{-2}	2.03×10^{-3}	4.38×10^{-28}
		dist	-3.08×10^{-2}	1.95×10^{-3}	4.97×10^{-56}

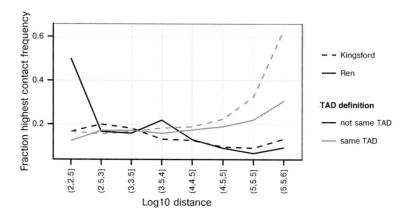

Figure 3.8: Gene pairs with strong binding dominate in TADs only for large distances. Gene pairs in the same distance bin were binned into 6 bins of equal size according to relative contact frequency as in Fig. 3.7b. The fraction of gene pairs in the bin of highest contact frequency is shown for four separate groups, separated according to TAD membership (same TAD: gray; not same TAD: black) and according to the TAD definition (Kingsford definition: broken line, Ren definition: solid line).

the most consistent impact of TADs on correlation for distances larger than 100 kb (Fig. 3.7a). This analysis also shows that the TAD definition proposed by Ren and coworkers induces less enrichment of strong contacts at long distances than the TAD definition proposed by Kingsford and coworkers. This lack of enrichment explains the weaker effect of TAD membership on correlation for the Ren definition. It also highlights that we need improved high-resolution chromatin conformation data and an improved understanding of what defines TAD boundaries for a definite assessment of the impact of TADs on co-expression.

We noticed that in each distance bin, average distances of gene pairs lying in the same TAD are shifted towards smaller distances compared to gene pairs not in the same TAD. The effect of increased correlation in TADs or in higher contact frequency bins can still be clearly observed, though (Fig. 3.7a). However, this bias may influence the statistical significance of the shift in the correlation distribution observed in Fig. 3.6b. To address this issue, we fitted linear models that allow for a modulation of the distance-dependent correlation decrease by TAD membership or contact frequency (Tab. 3.1).

Asking if TAD membership significantly modulates the decrease of correlation with distance, we first set up a linear model with a TAD membership dummy variable that allows for separate coefficients of the distance term. We found that the distance coefficient is less negative for genes in TADs than for other genes, though the error intervals for the two distance coefficients overlap. We obtained clearer results in the model containing an additive dummy variable for TAD membership and a distance effect. Here, the TAD membership coefficient is significant ($p < 10^{-7}$). Finally, we tested a model containing additive effects from the chromatin contact frequency of individual genes and genomic distance. We found that chromatin contacts increase correlation in a highly significant manner ($p < 10^{-27}$).

3.5 Expression fluctuations of neighboring loci are reduced in domains

As an independent test for the co-expression effect induced by TADs, we analyzed data from an experiment that probed the influence of the local environment of a gene on expression. In this experiment, reporters were first integrated at random positions in the genome of mESC and then their location and expression was determined (Akhtar et al., 2013). These reporters contain the same promoter and are thus perfectly suited for disentangling gene specific effects from purely context dependent effects. We reasoned that the co-expression induced by TADs should result in reduced expression variation among reporters lying in the same TAD compared to reporters not contained in the same TAD. To test this hypothesis, we compared the standard deviation of reporter expression in TADs to the standard deviation of reporter expression in random domains.

We obtained reporter expression generated by Akhtar et al. (2013) for two different promoter constructs, the mPGK promoter and the tet-Off promoter. Of these two constructs, the tet-Off promoter can be induced with doxycycline. The doxycycline concentrations used were 100ng, 0.1ng, 0.01ng and 0ng. We noticed that the standard deviation of the expression of all reporters was highest for the highest repression of the tet-Off promoter (Fig. 3.9a). This increased fluctuation makes these experiments more useful for detecting factors influencing the standard deviation. Consequently, for the tet-Off system, we computed for each reporter the average expression at doxycycline concentrations of 100ng and 0.1ng and compared the results with those obtained from the mPGK promoter data set (For further details, see Materials and Methods, Sec. 6.3, page 137).

In order to determine the effect of TADs, we compared the mean standard deviation of the reporter expression in TADs to a suitable null distribution. We

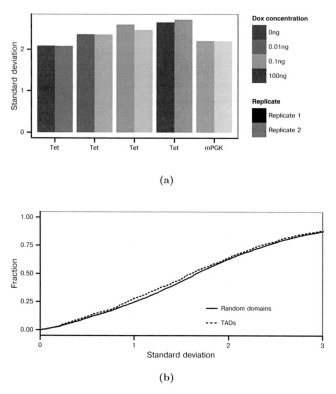

(a)

(b)

Figure 3.9: Reporter genes lying in the same TAD show reduced variability in expression compared to reporters lying in the same random domain. (a) Lowly induced tet-Off promoters lead to the highest variability in reporter gene expression. Standard deviation of expression for all reporter genes for the two reporter systems tet-Off (different doxycycline inductions) and mPGK used in (Akhtar et al., 2013). (b) The cumulated density function (cdf) for the reporter expression standard deviation in TADs is compared to the cdf for random domains. The Kingsford TAD definition and the tet-Off reporter data set are used. Note that not all data is shown so that differences in the distributions become better visible.

Table 3.2: Statistical significance of the reduced standard deviation of randomly integrated reporters in TADs. The indicated p-values were computed by comparing the distributions of the expression standard deviation for the indicated domains with random domains using Wilcoxon's rank-sum test.

TAD definition	Construct	p-value
Kingsford	Tet	0.035
Kingsford	mPGK	0.910
Ren	Tet	0.028
Ren	mPGK	0.032

computed the null distribution in two steps. First, we constructed random contiguous regions of the genome that have the same size and gap size distribution as the TAD definition used. Then we computed the average standard deviation for each partitioning of the genome in random domains. The null distribution was obtained from the standard deviations computed for 1000 reshufflings of domains.

We tested whether the standard deviation is reduced in TADs compared to random domains for the two alternative TAD definitions and the two reporter systems with a one-sided Wilcoxon rank-sum test. The resulting p-values listed in Tab. 3.2 show that we find a significantly lower standard deviation in TADs in all cases but the Kingsford TAD definition with the mPGK reporter system. Judging from the comparison of cumulated density functions for the Kingsford TAD definition and the tet-Off reporters (Fig. 3.9b), we see that the standard deviation shift is small compared to the width of the distributions. This may to some degree be caused by the fact that after filtering out zero expression values (for which the logarithm can not be computed) the median reporter distance is quite large with about 150 kb for the tet-Off system and about 65 kb for the mPGK system. Comparing this with the median size of 160 kb for the TADs defined by Kingsford one can see that many TADs will not contain two or more reporters, leading to poor statistics.

3.6 Proteins encoded in the same domain preferentially interact

It is widely accepted that co-expression of genes is a good proxy for shared function (Allocco et al., 2004; Michalak, 2008). So far our analysis of the functional role of TADs was exclusively restricted to expression data. The advantage of gene expression data is that high-throughput assays for this kind of data are highly developed, so that co-expression can be quantified with relatively low uncertainty. However, it is unclear what level of co-expression between two genes is needed to deduce a functional relationship. A certain fraction of co-expression between neighboring genes will be a byproduct of shared, accidental influences. Consequently, additional functional data should be used to assess the relevance of TADs.

Possible sources of such data are GO anotations, pathway annotations, complex membership and protein-protein interaction (PPI) data. In the following we will use PPI data as an additional source for functional data. Unlike data on complex membership and pathway annotations, PPI databases contain enough interactions to statistically assess a weak effect. Compared to shared GO annotations, PPI data has the advantage that there is greater consensus on the functional relevance of this kind of data.

As a source for PPI data we used StringDB (Mering et al., 2005; Franceschini et al., 2013). StringDB has the advantage of providing a score for a large number of gene pairs and being in continuous development. The StringDB protein-protein interaction database contains evidence from different sources:

- validated small-scale interactions, protein complexes, and annotated pathways (in the following: database)

- association in publications recovered via textmining (in the following: textmining)

- co-expression

- high-throughput yeast two-hybrid screens (in the following: experiment)

- conserved genomic neighborhood

- and phylogenetic co-occurence.

To avoid circular or redundant reasoning, we only used the following scores: database, experiment and textmining. Co-Expression scores would be redundant since we already analysed the association between TAD membership and co-expression. Scores based on genomic neighborhood might lead to circular reasoning if genes in TADs are more likely to remain in their genomic context over the course of evolution.

For each gene pair on the same chromosome we determined the mean of the database, experiment and textmining score provided by StringDB for this pair. Gene pairs with zero mean score were omitted from the analysis and the resulting mean was logarithmized. It has to be noted that the evidence from StringDB for gene pairs from the same chromosome is dominated by textmining-sourced evidence. We found 17591 unique gene pairs located on the same chromosome at a distance of less than 1 Mb with non-zero mean score. Of these, 15260 had a non-zero textmining score (87 %), 3184 had a non-zero database score (18 %) and 984 had a non-zero experiment score (6 %).

To determine whether proteins coded by genes in the same TAD preferentially interact, we adopted the following strategy. We selected a focal gene for which we retrieved the scores of all interaction partners. These scores were then normalized by computing the rank and dividing by the number of interaction partners. In order to determine the dependence of the interaction probability on genomic distance, we analyzed the distribution of normalized scores for all focal genes. We did this by binning interaction partners according to distance from the focal gene and TAD membership with the focal gene. For the TAD annotation, we used the Kingsford TAD definition which exhibited a more consistent association with co-expression.

The resulting normalized interaction score distribution was compared for all focal genes with at least 4, 5, 6 and 7 interactions (Fig. 3.10). We used this comparison because on the one hand relative effects can be better analyzed for focal genes with many interaction partners but on the other hand a stricter cutoff leads to poorer statistics. Nevertheless, we observed few differences for different interaction number cutoffs. We consistently detected a significant increase of the interaction probability in TADs in the distance bins spanning the range from 50 kb to 500 kb. For distances below 50 kb, we observed a consistent increase only for higher cutoffs (at least 6 or 7 interaction partners).

In each distance bin, the normalized interaction score fluctuates considerably around the median. To better compare the dependence of the score on distance, we also plotted the mean normalized interaction score in equi-distant logarithmic distance bins (Fig. 3.11). We observed that independent of TAD membership, interaction scores for partners at large distances are smaller than for partners at small distances. However, for interaction partners in the same TAD we found a non-monotonic dependence on distance. To check whether this is a property unique to TADs we recomputed the distance behaviour of interaction scores for random domains. As in the previous section, each random domain partitioning was constructed such that it exhibits the same domain and gap size distribution as the TAD partitioning of the genome. For random domains we find an identical, non-monotonic dependence of interaction scores on distance independent of domain membership of the interaction partner (Fig. 3.11). Thus the rapid decrease of the

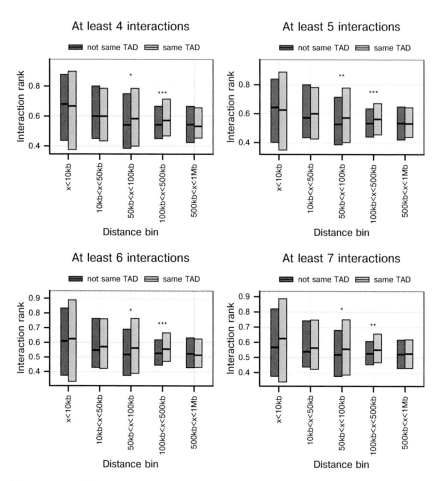

Figure 3.10: Protein-protein interaction data from the String database indicates that binding preference of a gene is shifted to more distant genes if both genes lie in the same TAD. For each gene, the interaction rank of its binding partners is computed from the String database score. Interaction partners are binned according to distance and TAD membership, where the Kingsford definition of TAD boundaries was used. Boxes indicate the interquartile range of the interaction partner rank. From top left to bottom right, panels show distributions of ranks for genes with at least 4, 5, 6 and 7 binding partners. Significantly different distributions are marked with an asterisk, with (∗), (∗∗), (∗∗∗) corresponding to $p < 0.05$, $p < 0.01$ and $p < 0.001$, respectively (Wilcoxon rank-sum test).

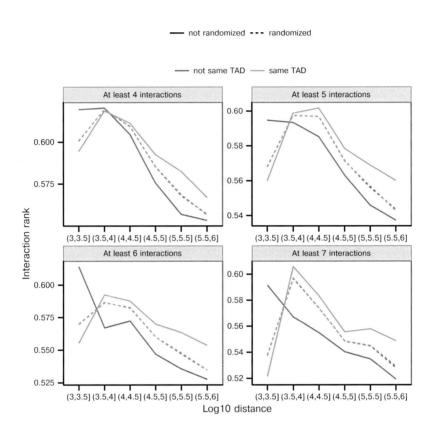

Figure 3.11: Mean interaction rank behavior shows distinct profile with respect to gene pair TAD membership (Kingsford definition) that is lost for random domains. Gene pairs are binned with respect to equidistant logarithmic distance intervals. For each gene, the interaction rank of its binding partners is computed from the String database score and average scores are computed separately for interaction partners in the same TAD (blue line) and interaction partners not in the same TAD (red line). The same computation is repeated for 250 reshufflings of random domains with same size and gap size distribution as the real TADs (broken lines). Different panels contain data with the minimum number of interactions per gene indicated, ranging from a minimum of 4 to a minimum of 7 interaction partners.

interaction score with distance is a special feature of interaction pairs that are not in the same TAD or do not lie in a TAD at all. This implies that for these interaction partners, there is a strong preference for nearby partners.

We noted above that most of the predicted intra-chromosome interactions in StringDB are based on textmining evidence. To check whether interactions based on other types of evidence lead to qualitatively different results, we performed the above analysis also using only evidence labeled as database-sourced. This is evidence coming from small scale experiments and curated knowledge about protein complexes and pathways (Mering et al., 2005). The resulting plot (Fig. 7.3, page 145) supports the observation that interactions are more likely among genes in the same TAD than among other genes. In contrast to the results based mainly on textmining evidence, we observed no rapid decrease of interaction scores for partners that are not in the same TAD.

Taken together, our analysis shows that for long distance interactions, a given gene is more likely to interact with a gene in the same TAD than with genes in other TADs or genes that are not in a TAD. For PPI data dominated by textmining evidence, we also observed a rapid decrease for interaction scores among gene pairs that are not in the same TAD.

3.7 Discussion and outlook

The question to which extent eukaryotic gene order is non-random has been extensively studied in the past (Hurst et al., 2004). Research into long-range chromatin contacts and chromatin conformation in general has significant impact on this questions because it has redefined the notion of close and distant genes. We showed that strong chromatin contacts are associated with correlated expression, allowing coordinated expression over genomic distances in the range of a few Mb. We first investigated a region on the X chromosome around the X inactivation center. There, we found that the block like structure of chromatin contacts is associated with coordinated expression within the blocks. Among these blocks, called TADs, we found two which align with functional groups of Xist activating or Xist repressing genes. In a second step, we showed that coordinated expression in TADs can be observed genome-wide, though the strength of the co-expression effect depends on the details of the domain definition. These findings are based on gene expression data of early differentiation in mESC. As a complementary data set, we analyzed the impact of chromatin conformation on the expression of randomly integrated reporter genes. We found that differences in expression are significantly reduced in TADs compared to a null model of random domains. Additional evidence for a functional relationship between genes in the same TAD was shown to come from protein-protein interaction data. Proteins encoded by distant genes from the same TAD have a higher tendency to interact than proteins encoded by distant genes not from the same TAD.

Our analysis profits from the advances in the field of 'C'-techniques, making use of 5C and Hi-C data which yields chromatin contacts at a resolution of the order of 10 kb. A similar analysis (Woo et al., 2010) using less well resolved human Hi-C data (resolution of 1 Mb) also showed that contact frequency and co-expression are correlated but crucially failed to control for genomic distance. Interestingly, this study found that intervening CTCF binding sites reduce the co-expression of a gene pair. It was later established that binding sites of the insulating factor CTCF are enriched at TAD boundaries (Dixon et al., 2012; Filippova et al., 2014).

If TADs constitute a general principle for coordinated gene expression, their impact should not only be detected in cell differentiation or development but also in perturbation experiments. Such an experiment was employed by a study (Dily et al., 2014) investigating changes in gene expression after stimulation of a human cell line with a hormone. The authors found that the number of TADs containing only up- or downregulated genes is significantly higher for the true gene order than for a randomized gene order.

Our analysis shows that TADs and chromatin contacts in general are associated with significantly increased co-expression. To our knowledge this is the only study finding such an effect while controlling for genomic distance. Our approach makes use of the fundamental change in expression state associated with stem cell differentiation, which involves a considerable fraction of all genes (over 5%). Still, though highly significant, the increase in correlation in TADs is small, raising the question about the functional significance of TADs as a principle of genome organization. For this reason we investigated whether additional evidence for shared function in TADs comes from protein-protein interaction data. Our observation that distant genes (distance > 50 kb) tend to interact with genes from the same TAD is supported by an analysis into the 3d distance of genes in functional groups (Thévenin et al., 2014). In this study, it was shown that genes that form a complex or genes that are members of the same pathway are closer in space than expected from a random order of genes on the genome.

Still, too further assess the functional significance of TADs, our approach should be complemented by using data from different tissues. Genes from many TADs may only be active in certain tissues, making the regulatory impact of these TADs impossible to detect in our data. This might explain the rather weak effect of TADs on co-expression in the differentiation time series.

The presented investigation should also be extended in other directions. Apart from looking at the average effect of coordinated expression associated with TADs, our genome-wide analysis can also be used to prioritize certain TADs for a detailed analysis. This detailed analysis can then be used to infer hypotheses for principles of gene regulation mediated by TADs. Another important aspect is to understand the mechanisms that give rise to TADs or allow their reconfiguration. This question is related to characterizing the architectural proteins that stabilize the chromatin conformation and the dynamics of these proteins (Gómez-Díaz and Corces, 2014).

Finally, TADs have been found to align with other chromatin features such as histone marks (Nora et al., 2013), raising the question whether TADs are cause or consequence of other features. An important feature of chromatin are the lamina associated domains (LADs) that are associated with attenuation of transcription and control of replication timing (Wit and Steensel, 2009; Yaffe et al., 2010). A study by Akhtar et al. (2013) found LADs to be aligned to domains of expressed or silent integrated reporters in mESC . The authors of this study claim that there is no alignment of TADs and LADs, whereas we found an effect of TAD membership on reporter expression using the same data set.

The lack of alignment between TADs and LADs has recently been analysed by Pope et al. (2014), confirming a model that had been proposed earlier (Nora et al., 2013). This model posits that TADs are stable units which can dynamically attach to or detach from the nuclear lamina. Thus, LAD borders should always be TAD borders but not vice versa. This relationship between LADs and TADs supports the notion that TADs are fundamental units of gene regulation (Denholtz et al., 2013).

4 Uncovering regulation of individual genes by transcription factors in mESCs

4.1 Introduction

In the previous two chapters we have seen examples of regulatory processes that operate on groups of genes. In this chapter we are interested in the unique expression profile of individual genes. This expression profile is determined by regulatory elements that control a gene and the state of the local chromatin that modulates this control. Regulatory elements may be roughly grouped into distal elements such as enhancers and silencers and proximal elements that are made up by the core promoter and elements closer than 1 kb to the promoter (Maston et al., 2006). The important role of distal elements for gene regulation is increasingly appreciated in the literature. However, the precise characterization of distal elements remains a challenge and they are less well understood than proximal elements. For this reason, we will focus in this chapter on the proximal regulatory elements.

The core promoter and the promoter proximal elements may contain binding sites for transcription factors (TFs) that activate or repress transcription. In principle, a promoter can be unique for a given gene, giving rise to a huge number of combinatorial binding possibilities. Collecting the information about the TFs that interact with each promoter, we arrive at the topology of the gene regulatory network (GRN). This network is thought to determine a large part of gene expression, yielding important information about the relationship among different genes and how they interact to drive cellular programs. In mouse embryonic stem cells (mESCs) in particular, we are interested in the temporal sequence of differentiation events and the sensitivity of the differentiation process with regard to perturbations of individual genes.

The question guiding this chapter is thus: to which degree is it possible to derive the topology of the GRN from the expression patterns we observe during differentiation? Put more concretely, we will proceed in two steps. First we will reconstruct the topology of the gene regulatory network governing the differentiation process from transcriptome data. Second, we will ask how well this reconstructed network is in agreement with mechanistic insights into the biological network. Among others, we will draw these mechanistic insights from assays measuring TF-DNA binding and transcriptome data from TF perturbation experiments. These two classes of experiments are the two main sources for our current understanding of gene regulatory networks.

How does the presented approach differ from other network reconstruction approaches? A large number of approaches to the problem of inferring gene regulatory networks from high throughput data have been developed in the last 20 years. Many of these approaches have been systematically benchmarked at different stages of the DREAM project (Stolovitzky et al., 2007). However, it is unclear how the performance of different approaches in the single cell organisms *E. coli* and *S. cerevisiae* used in the DREAM project compares to mammalian cells and more specifically to differentiating mESC. Cellular differentiation is unique to multicellular organisms and requires sophisticated gene regulation. The regulatory program governing cellular differentiation has to allow for propagation of the undifferentiated, pluripotent state as well as the irreversible commitment to certain cell lineages upon certain developmental cues. This combination of flexibility and stability (Reik, 2007) has shaped the wiring of genes responsible for orchestrating the differentiation process. As cellular differentiation is unique to multicellular organisms, so should this wiring be unique to these organisms. Consequently, we can not infer the reconstruction quality of the network governing differentiation in mESC from benchmarks of the DREAM project.

In order to reconstruct gene regulatory networks from microarray data, we have to make the assumption that the rate of change of a gene's mRNA concentration is a function of the mRNA concentration of all other genes (Margolin and Califano, 2007). This amounts to two implicit assumptions. First, the protein concentration of each gene is proportional to its mRNA concentration and second, proteins do not have to be post-translationally modified in order to be regulatory active. If this is satisfied, we can write the dynamical equations governing the mRNA expression dynamics as

$$\dot{x}_i = f_i \left(\{x_j\}_{j \in \{1,\dots,n\}} \right), \tag{4.1}$$

where $f : \mathbb{R}^n \to \mathbb{R}$ encodes the functional form of the gene induction and x_i are the gene concentrations. In the following we will furthermore assume that only transcription factors can regulate other genes. Numbering genes such that the first m are the transcription factors, the argument of f becomes x_j with $j \in \{1, \dots, m\}$

Around a steady state p, we can linearly expand the right hand side to obtain

$$0 = J_f(p)(x - p) + o(\|x - p\|), \tag{4.2}$$

where $J_f(p)$ is the Jacobian of f at the point p. Assuming that degradation rates of x_i only depend on the concentration of x_i in a linear manner and that no self activation takes place, we can write

$$\sum_{j \neq i} A_{ij}(p)\Delta x_j - \lambda_i \Delta x_i = 0, \tag{4.3}$$

where $\Delta x_i = (x_i - p_i)$. If we had expression measurements around one single steady state, we could determine the A_{ij} by regression or other methods. However, as we would like to use expression data coming from all stages of differentiation, we have to make the strong assumption that $A_{ij}(p)$ does actually not depend on the steady state p. We then obtain the simple equation

$$\Delta x_i = \sum_{j \neq i} \frac{A_{ij}}{\lambda_i} \Delta x_j, \tag{4.4}$$

indicating that a change in x_j leads to a change proportional to A_{ij} in x_i. The A_{ij} thus measure the influence of x_j on x_i, reflecting the topology of the network. The goal of the reconstruction of gene regulatory networks is to identify the sign of the A_{ij}, corresponding to 1 for activatory links and -1 for inhibitory links.

We can now precisely state the problem that shall be investigated in this chapter. Given a set of N steady state transcriptome samples collected in a matrix B, which algorithm predicts sign(A_{ij}) in such a way that it has the highest overlap with biologically meaningful gold standards. Which gold standards are considered biologically meaningful will be discussed later.

Now that we have formulated the problem, we can discuss the progress that has been made so far in solving it. Putting it more generally, what is the state of the art of network reconstruction in ESC? To our knowledge, no reconstruction effort making use of large scale public transcriptome data has been comprehensively benchmarked. Benchmarks have so far focused on simulated data or data from *E. coli*. Allen and coworkers (Allen et al., 2012) study the influence of sample size on reconstruction quality in a simulated network and compare reconstruction algorithms also on over 500 microarray samples from *E. coli*. Few studies have benchmarked reconstruction algorithms on mammalian data sets. One study

(Bansal et al., 2007) applies clustering, information theoretic, Bayesian network inference and regression based algorithms to multiple cells/organisms, among them human B-cells, using a gold standard from Basso et al. (2005). In this comparison, the ARACNE algorithm performs best on the B-cell data. Song et al. (2012) make use of mammalian data sets but use only GO enrichment as a performance metric.

The literature dealing with reconstruction of ESC-specific gene regulatory networks from large scale transcriptomics data has mainly avoided the topic of reconstruction quality. Two exceptions are found in the works of Cegli et al. (2013) and Cahan et al. (2014). Cegli et al. used ARACNE on 171 microarrays to reconstruct a gene regulatory network in mESC. Reconstruction quality is assessed using Reactome and the Escape database (Logof data set). Cahan et al. used around 4000 samples from different mouse tissues to train a gene regulatory network using the CLR algorithm. As quality benchmark they computed the improvement over random for the area under the precision-recall-curve (AUPR) for different data sets: the ChIP-Chip data in the Escape database, a data set measuring differential expression after overexpression and the Encode ChIP-seq data. However, this work does not include any comparison of reconstruction methods, making the improvement over random values hard to interpret.

There exist reconstruction attempts for the gene regulatory network in mESC (Kim et al., 2008; Dowell et al., 2013), that use ChIP-seq data additionally or exclusively for training and are thus not directly comparable to approaches using only transcriptome data. Finally, some reconstruction efforts aim at only reconstructing a core pluripotency network containing a couple of nodes, using time resolved data (Anchang et al., 2009) or culture condition dependent expression (Dunn et al., 2014).

Here, we use a large collection of over 1000 public transcriptome samples to infer the gene regulatory network in mESC using a set of existing reconstruction algorithms (Fig. 4.1). We compared the predictions of these algorithms with biological evidence for functional targets of TFs. This evidence comprises ChIP-seq data as well as transcriptome data that identifies differential expression after transcription factor perturbation. The individual algorithms were then ranked by their average performance across the different types of evidence. This ranking showed that only algorithms that employ strategies to eliminate indirect links achieve a high rank. Finally, the best-performing algorithms are used to reconstruct the TF-TF regulatory network. The topologies of the reconstructed networks are compared to the topology of a curated, literature-based network. The comparison of the topologies showed that the reconstruction strategies employed by different algorithms strongly influence the topological properties of the reconstructed network.

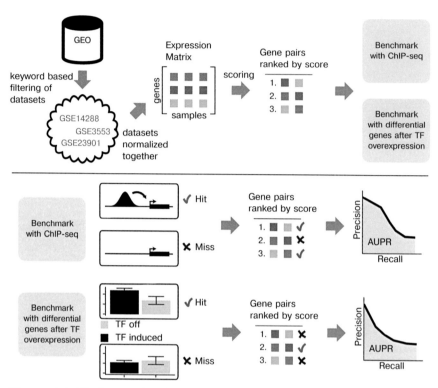

Figure 4.1: Pipeline used in this chapter for the benchmarking of network prediction algorithms. Upper panel: Microarray samples from a fixed Affymetrix mouse array type are downloaded from the GEO database that are identified by keywords related to embryonic stem cells. All samples are normalized together, yielding the gene expression matrix. Different algorithms that predict interactions based on co-expression are then applied to the expression matrix to obtain a list of gene pairs ranked by their interaction score. Since we investigate gene regulation by TFs, only TF-gene pairs are retained. The quality of predictions is evaluated using two kinds of evidence for direct interaction, ChIP-seq and differential expression after TF perturbation. Lower panel: TF-gene pairs are classified as hits and miss according to ChIP-seq and TF perturbation evidence. For the ChIP-seq gold standard, a TF-gene pair is classified as hit when the TF binds close enough to the promoter of the gene. For the perturbation gold standard, genes with a significant fold change upon TF perturbation are classified as targets of the respective TF. From the rank of the hit and miss TF-gene pairs in the list ranked by interaction score, the precision recall curve is calculated. The area under the precision recall curve (AUPR) serves as quality benchmark of the network predictions.

4.2 Statistical measures of association used for network prediction

A vast number of measures of associations or algorithms for reconstructing gene regulatory networks in an unsupervised fashion from expression data have been proposed in the literature (Marbach et al., 2012). It is impractical to assemble an exhaustive set of measures for benchmarking. Instead we will make use of a classification of measures and pick representatives for each class. Often measures are classified by the mathematical theory on which they are founded (de Jong, 2002; He et al., 2009). Focusing on co-expression patterns, we adopted a classification that has been proposed by Santos et al. (2013). It is based on a classification of possible relationships between non-independent random variables. These relationships can be classified in a hierarchical way with the most special case given by a linear relationship and the most general case given by a non-functional relationship. The complete hierarchy of classes is non-functional, functional non-monotonic, functional monotonic non-linear and linear (Fig. 4.2).

Apart from statistical measures of association where the score for a gene pair only depends on the expression vectors of both genes, we also included more complicated algorithms that try to identify direct relationships between random variables. This is done by either controlling for the influence of other random variables on the relationship between two random variables (partial correlation) or by comparing the strength of association between different pairs of variables (e.g. ARACNE). Combining the ability to identify direct relationships with the hierarchical classification of relationships, we arrive at Tab. 4.1, describing the properties of the different scores employed by us.

In our analysis we included Pearson correlation, Spearman correlation, mutual information, ARACNE (Margolin et al., 2006), MRNET (Meyer et al., 2007), CLR (Faith et al., 2007) and partial correlation. For each of these measures of association, one can check the ability to reject the null hypothesis of independence for relationships falling into one of the co-expression classes discussed above. Pearson correlation is maximal (or minimal) when all observations lie on a line, indicating that it performs best on linear relationships. Spearman correlation is Pearson correlation computed on ranked data, implying that the prerequisite of a linear relationship is relaxed to a monotonic relationship. Mutual information for two random variables X and Y is zero if and only if X and Y are independent (Santos

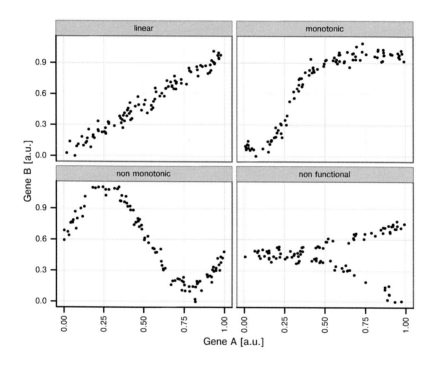

Figure 4.2: Overview of the different classes of co-expression patterns. The patterns are arranged by increasing generality from linear over functional monotonic and functional non-monotonic to non-functional.

Table 4.1: Classification of scores with the help of the scheme proposed in (Santos et al., 2013). Scores are classified by their ability to detect co-expression relationships with the indicated properties (n-lin: non-linear, n-mon: non-monotonic, n-func: non-functional, rem indir: removes indirect links).

score	n-lin	n-mon	n-func	rem indir
Pearson correlation	no	no	no	no
Spearman correlation	yes	no	no	no
Mutual information	yes	yes	yes	no
ARACNE	yes	yes	yes	yes
CLR	yes	yes	yes	yes
MRNET	yes	yes	yes	yes
Partial correlation	no	no	no	yes

et al., 2013). Put another way, mutual information can detect the most general relationships that we termed non-functional. As ARACNE, MRNET and CLR are all based on mutual information, they inherit the property of being able to identify non-functional relationships. Partial correlation for two random variables X and Y is defined as the Pearson correlation between residuals r_X and r_Y from the linear regression of X and Y on the controlling variables. Consequently partial correlation is limited to the same class of relationships as Pearson correlation.

How relevant is it for reconstruction if a measure of association can identify very general co-expression patterns? Non-linearity necessarily arises because all gene concentrations in a cell have to saturate eventually. Furthermore, inducible synthesis has often been found to be accurately described by Hill functions (Glass and Kauffman, 1973) that are widely used in modeling gene regulation (de Jong, 2002). Non-monotonic and non-functional relationships are in principle possible, for instance when cofactors modulate the regulation by the primary transcription factor. In this sense we can think of non-functional co-expression patterns as two dimensional projections of a higher dimensional state space, involving one or more cofactors. However, many non-functional relationships may not be biologically meaningful, so the ability to detect such relationships may even reduce the quality of the predictions.

All algorithms take a $n \times m$ matrix as input, where n is the number of genes and m is the number of (microarray) samples. The output is an $n \times n$ matrix that contains an association score for each gene pair. We set the diagonal entries to zero if the algorithm does not do this automatically. Diagonal entries have to be set zero since self-interaction can not be inferred from expression data. All mutual information based scores, mutual information, ARACNE, CLR and MRNET are computed with the R package parmigene (Sales and Romualdi, 2011). The ARACNE algorithm is the only one which needs an input parameter. This parameter τ determines how aggressive the algorithm tries to prune indirect links. The value $\tau = 1$ corresponds to no pruning, while for $\tau = 0$, the weakest link between three mutually connected nodes will always be deleted. In this work, we employed $\tau = 0.15$ (denoted by ARACNE$_{15}$) and $\tau = 0.5$ (denoted by ARACNE$_{50}$) corresponding to the preconfigured standard value and very weak pruning, respectively.

For partial correlation we used the R package parcor (Krämer et al., 2009), that offers multiple strategies to estimate the partial correlation matrix. We chose a sparse (most entries of the partial correlation matrix are set to zero) and a non-sparse method for estimating partial correlation. The non-sparse method is based on partial least squares (pcor.pls) and the sparse method on a lasso regression (pcor.lasso) (Krämer et al., 2009).

4.3 Gold standards for determining direct TF-gene interactions

A considerable part of the network reconstruction literature deals with relatively simple model organisms such as *E. coli* or *S. cerevisiae*. A main reason for testing reconstruction on data from *E. coli* is the existence of a generally accepted gold standard containing curated gene regulatory interactions given by RegulonDB (Huerta et al., 1998; Salgado et al., 2006). The situation for mammalian organisms is more complex, where no one gold standard is used on a regular basis. Thus, we have to survey the existing gold standards, each of which has its advantages and disadvantages.

Among the gold standards used in the network reconstruction literature in general are curated interactions (Marbach et al., 2012; Zoppoli et al., 2010; Zhang et al., 2012; Kaderali and Radde, 2008; Bansal et al., 2007; Castelo and Roverato, 2009; Cegli et al., 2013; Ciofani et al., 2012; Faith et al., 2007; Li et al., 2013), binding data based on ChIP-chip or ChIP-seq (Marbach et al., 2012; Cahan et al., 2014; Haynes et al., 2013; Honkela et al., 2010; Margolin et al., 2006; Roy et al., 2013; Jang et al., 2013; Joshi et al., 2014), binding motifs (Marbach et al., 2012; Haynes et al., 2013), differential expression upon TF overexpression or knockdown (Cegli et al., 2013; Cahan et al., 2014; Jang et al., 2013), enrichment of GO terms(Song et al., 2012; Castelo and Roverato, 2009) or even rationally designed biological

Table 4.2: Parameters describing the data sets on which the gold standards ChIP-seq, LoF (TF loss-of-function), Kd (TF knockdown) Overexpression (TF overexpression) for benchmarking network reconstruction algorithms are based. The first two parameters, FDR and minimum fold change describe the stringency of the cutoff for differential expression. The last row (Fraction of possible) indicates the fraction of links in the network compared to a fully connected network with the same number of nodes.

	ChIP-seq	LoF	Kd	Overexpression
FDR	NA	mixed	0.05	0.001
minimum fold change	NA	mixed	2	2
# Interactions	4000	17077	6709	61587
# TFs	16	19	41	90
Mean interactions TF	250	898.79	163.63	684.30
Fraction of possible	0.0127	0.0458	0.0083	0.0349

networks (Zoppoli et al., 2010). A large part of the reconstruction literature also makes use of simulated gene regulatory networks (Zhu et al., 2012; Zhang et al., 2012; Yuan et al., 2011; Schäfer and Strimmer, 2005; Kaderali and Radde, 2008; Krämer et al., 2009; Liang et al., 2012; Marbach et al., 2010; Margolin et al., 2006).

The conclusions we can draw from simulated networks for the reconstruction of mammalian gene regulatory networks are limited, because there is still considerable uncertainty about how transcript levels are actually regulated. This ranges from questions regarding the extent of noise in the transcript levels that is tolerable for a cell over the complexity or interconnectedness of typical gene regulatory networks to the specificity of transcription factor - DNA interaction. Simulations make explicit or implicit assumptions about these problems. In particular, simulations have to adopt certain distributions of expression values for regulating factors and certain shapes of induction curves for regulated genes. These assumptions are hard to verify and might be quite unrealistic in embryonic stem cells.

Consequently, we believe that the quality of a network reconstruction algorithm in embryonic stem cells should be judged by using gold standards that are based on experiments in these cells. We chose the gold standards for our benchmark based on two criteria. First, they have to identify direct interactions that are also functionally relevant. Second, the available data for individual gold standards should cover a reasonable number of TFs. No one gold standard satisifies all these criteria.

Thus, we chose as gold standard used for our benchmark a combination of two different types of evidence which are available for a sufficient number of transcription factors: a collection of TF ChIP-seq data sets (Mouse ES Cell ChIP-Seq Compendium), a collection of TF knockdown experiments (described as loss-of-function experiments in the database, LoF) gathered in the Escape database (Xu et al., 2013), a set of TF knockdown experiments followed by transcriptome analysis (Kd) (Nishiyama et al., 2013) and a set of TF overexpression experiments followed by transcriptome analysis (Overexpression) (Nishiyama et al., 2009; Correa-Cerro et al., 2011) (See Materials and Methods, Sec. 6.4, page 137). Each gold standard was restricted to transcription factors using transcription factor annotation from the Animal Transcription Factor Database (ATFDB) (Zhang et al., 2011). The combination of ChIP-seq data and transcriptome data after TF perturbation is particularly suited because these two types of assays are complementary. While ChIP-seq identifies direct interactions, differential expression after perturbation is also assumed to be a necessary precondition for a functional interaction (Cusanovich et al., 2014).

For the different gold standards true interactions were identified as follows. For the ChIP-seq gold standard, the input data consisted in binding peak locations and peak height scores derived from the raw sequencing data. For each gene, an association score was calculated that is given by a sum of the scores of nearby peaks weighted by their distance (For details, see Materials and Methods, Sec. 6.4, page 137). For each TF, the top 250 genes with the highest association score were defined as true targets. For the three TF perturbation data sets, true interactions were defined based on differential expression. Two parameters define the genes that were taken to be differentially expressed, minimum fold change and false discovery rate (FDR). For the LoF and Kd data set, we used the provided lists of differential genes in the Escape database and in Nishiyama et al. (2013), respectively. The cutoffs used to generate these lists were mixed in the case of LoF data set because they are sourced from different publications. The Kd data set used FDR< 0.05 and a fold change > 2 as cutoff. For the Overexpression data set we used normalized expression data from Nishiyama et al. (2009) and Correa-Cerro et al. (2011) provided in the GEO database. We employed strict criteria (FDR< 0.001, fold change> 2) to infer the differential genes for each TF perturbation. An overview of the cutoff parameters and size of the network for all gold standards is given in Tab. 4.2.

It should be noted that of course all used gold standards have their drawbacks and do not necessarily identify direct, functional interactions. While ChIP-seq suffers

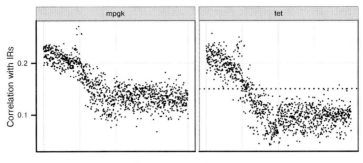

Figure 4.3: Position-dependent expression determined in stem cells can be used to classify samples into mESC/non-mESC samples. For each sample from the Mouse Gene St platform the Pearson correlation of gene expression and average integrated reporter (IR) expression is indicated by a point. In left panel reporters are based on the mPGK promoter construct, in the right panel reporters are based on the tet-Off promoter construct. Average coefficients are calculated by correlating the expression of each gene with the expression of the nearest reporter. Stemness is defined by the mean expression of Pou5f1, Sox2 and Nanog and samples are ordered with stemness decreasing from the left to the right. The stem cell cutoff used is indicated by a horizontal broken line.

from the identification of many false positives (Cusanovich et al., 2014; Spivakov, 2014), assays based on measuring differential expression are prone to identifying indirect interactions. A crucial parameter in forced induction or knockdown experiments is the time period which lies between induction/knockdown and the transcriptome measurement. To maximize the effect of the induction/knockdown, experimenters often wait for longer time periods and risk detecting many secondary expression effects.

4.4 How the transcriptome data was obtained

In order to make use of the large amount of transcriptome data sets available, we developed a simple data aggregation strategy. For one Affymetrix microarray platform, we downloaded raw sample data from the GEO database based on matching keywords in the abstract to a predefined list. We searched the abstracts of GEO database data sets for the following keywords: mESC, stem cell, stem cells, Oct4,

Sox2, Nanog, Pou5f1 and embryonic. This yielded 1194 samples for the Affymetrix Mouse Gene 1.0 ST Array (Gene St) platform. The raw cel files corresponding to these samples were downloaded and normalized together to obtain an expression matrix (See Fig. 4.1, page 82).

It is clear that neither will all samples from mESC be covered by this strategy nor can we ensure to obtain only mESC samples. To assess the composition of the downloaded samples, we compared their expression patterns with that of randomly integrated reporters (IRs) in mESC (Akhtar et al., 2013), already discussed in section 3.5. This comparison is based on the assumption that the position dependent expression exhibited by the reporters should be most predictive for gene expression in stem cells. As cells differentiate and chromatin is remodeled, the ability of reporter expression to predict gene expression should decrease. This can be quantitated simply by computing the correlation between genes and nearby reporters for each sample.

We started to test this approach by computing the mean expression of Nanog, Sox2, and Pou5f1 in each sample. Since it is well known that these three genes are highly expressed in stem cells and are all downregulated during differentiation, we used their mean expression to define a crude measure of stemness, i.e. closeness to the expression state of stem cells. We then ordered all samples by this stemness score and computed the correlation of the gene expression in each sample with that of the nearby reporters (For details, see Materials and Methods, Sec. 6.4, page 140).

The result of this analysis (Fig. 4.3) shows stemness defined by expression of Nanog, Sox2 and Pou5f1 and stemness defined by correlation with IR expression are consistent. Using correlation with IRs we can make a rough definition by defining samples with a correlation greater than 0.15 as coming from pluripotent stem cells. This definition has the advantage that it does not depend on the scale of the expression values obtained from microarray samples. We apply this definition to the correlation based on the tet-Off promoter construct since this shows the higher variance and thus can better discriminate between samples. Using this definition we find 28.4% stem cell samples among all samples from the Gene St platform.

The relatively low fraction of undifferentiated stem cell samples points towards a considerable heterogeneity in the assembled data sets. However, both samples of undifferentiated stem cells and of cells that have lost pluripotency are likely needed for reconstruction. This is because the presence of both leads to sufficient variance in the expression of pluripotency-associated TFs, which in turn allows to identify

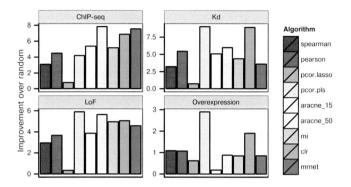

Figure 4.4: Improvement of the $AUPR_{0.05}$ over random predictions for networks predicted by individual scores. Improvements are shown for four gold standards based on ChIP-seq measurements (ChIP-seq), differential gene expression after TF knockdown (Kd), TF loss of function experiments (LoF) and differential gene expression upon TF overexpression (Overexpression). Algorithms used for the comparison are Spearman correlation (spearman), Pearson correlation (pearson), mutual information (mi), ARACNE (with cutoff parameter 0.15 and 0.5, respectively), CLR, MRNET and partial correlation in the pls (pcor.pls), lasso (pcor.lasso) implementation.

genes that co-vary with these TFs. For this reason, all samples from the data set will be kept for reconstruction.

4.5 Benchmark of network predictions: pruning determines success

Having introduced the different algorithms, the biological gold standards used to judge their performance and the transcriptome data the algorithms operate on, the last prerequisite for a benchmark consists in a score that quantifies the performance. In the following the area under the precision recall curve with a recall threshold of 5% will be used as a score, which will be abbreviated AUPR. The precision recall curve measures how the fraction of true predictions declines as we move down a list of ranked predictions from the top predictions to the bottom predictions.

Table 4.3: Average ranks of algorithms with respect to the four gold standards: ChIP-seq, LoF, Kd and Overexpression. Low ranks indicate comparatively bad performance, high ranks good performance. Algorithms used for the comparison are Spearman correlation (spearman), Pearson correlation (pearson), mutual information (mi), ARACNE (with cutoff parameter 0.15 and 0.5, respectively), CLR, MRNET and partial correlation in the pls (pcor.pls), lasso (pcor.lasso) implementation.

	Algorithm	Mean rank
1	pcor.lasso	1.25
2	spearman	3.25
3	aracne_15	4.00
4	mi	4.50
5	pearson	4.75
6	mrnet	5.00
7	aracne_50	7.25
8	pcor.pls	7.50
9	clr	7.50

For each algorithm, individual TF-target interactions were ranked by the absolute value of the corresponding score predicted for this pair. Together with the information from the gold standards, indicating which TF-target pair corresponds to a truly existing interaction, the precision recall curve can be computed. As a benchmark, the improvement of the AUPR over the theoretical random performance was calculated (For details, see Materials and Methods, Sec. 6.4, page 141). The improvement over random predictions for individual algorithms with respect to the different gold standards are shown in Fig. 4.4. The ChIP-seq and LoF gold standards are less sensitive to differences between the algorithms, as can be seen from the relative small performance differences of the top performing algorithms. In contrast, in the case of the Overexpression and Kd benchmark, one or two algorithms perform much better than the rest.

Looking at the individual algorithms, partial correlation in the pls implementation and CLR attain the highest average rank across gold standards of all algorithms (Tab. 4.3). Except for the ChIP-seq gold standard, pls-based partial correlation occupies the top rank for each gold standard. The two algorithms CLR and ARACNE with cutoff parameter $\tau=0.5$, both based on mutual information but using additional strategies for eliminating indirect links, occupy the top three ranks together with pcor.pls. Thus all three top performing algorithms employ a

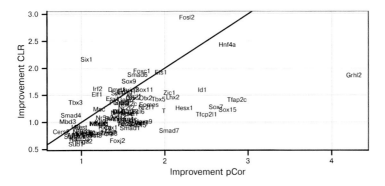

Figure 4.5: The two algorithms CLR and pcor.pls differ in the quality of predictions for individual TFs and yield complementary predictions. Scatter plot of the improvement of the $AUPR_{0.05}$ over random predictions for the CLR and pcor.pls algorithms. The predictions are benchmarked using the Overexpression gold standard.

strategy for pruning indirect links. On the other hand, being able to detect only linear relationships does not impact performance strongly. This is indicated by the fact that partial correlation is among the best performing algorithms and that Pearson correlation and mutual information attain a similar average rank. We concluded that reconstruction success of an algorithm is determined more by the pruning strategy used than by the co-expression patterns detected.

We were surprised the two implementations for computing partial correlation performed very differently. Indeed they occupy the opposite ends of the ranked list of algorithms. The top performance of the pls implementation is all the more surprising since it has the highest error for estimating the partial correlation matrix on synthetic data (Krämer et al., 2009).

We also measured the performance of different algorithms for each TF individually. For this comparison, we used the AUPR for the full precision recall curve (recall cutoff equal to 1) since the lower amount of predictions in this case compared to predicting the entire network would otherwise lead to higher fluctuations in the AUPR. The resulting improvements over random predictions are shown in Fig. 7.5, page 147.

We found the performance to vary considerably among TFs. Some of this variation may be due to the differing number of true targets between TFs in the three

Table 4.4: Improvement of the AUPR over random for different algorithms with respect to the gold standard given by the PluriNetWork at a recall of 5% ($AUPR_{0.05}$) and 100% ($AUPR_{1.00}$).

	pcor.pls	clr	aracne_50
Improvement $AUPR_{0.05}$	18.26	20.46	20.42
Improvement $AUPR_{1.00}$	2.77	3.3	3.18

perturbation gold standards (Kd, LoF, Overexpression). But also in the case of the ChIP-seq gold standard, where the random AUPR is the same for each TF because the number of targets was chosen to be fixed, we found differences. For the ChIP-seq gold standard and the CLR algorithm, the improvement over random ranges from about 1 to more than 3. This raises the question whether the performance of different algorithms is correlated among TFs. To check this, we computed the correlation of the improvement of the AUPR over random for individual TFs for the four different gold standards separately and plotted the average correlation matrix (Fig. 7.4, page 146). The average correlation matrix reveals that all scores except for the two partial correlation implementations and ARACNE with the default parameter τ=0.15 form a block with highly correlated performance. Inside this block, we can discern the a subgroup of mutual information derived scores, made up of the mutual information score itself, CLR, $ARACNE_{50}$ and MRNET.

The pcor.pls score is the only score that is both not correlated to the mutual information derived scores and performs well with respect to the overall improvement over random. That makes this score an interesting candidate for providing TF targets that complement the predictions of other well performing scores. This is especially obvious when comparing the performance of pcor.pls and the CLR algorithm for individual TFs determined on the overexpression gold standard (Fig 4.5). Here, we observed a lot of off-diagonal data points, indicating TFs for which predictions of one of the algorithms are superior to that of the other.

4.6 Predicted topologies of the TF-TF network differ strongly

We also used the top performing algorithms to predict the topology of the TF-TF gene regulatory network in differentiating mESC. With the topology of this network, several open questions can be addressed. Two central questions may serve as an example. Is there really a clear hierarchy between core pluripotency factors and ancillary pluripotency factors as proposed by a recent model (Dunn et al., 2014)? How are pluripotency factors coupled to lineage specific factors in order to regulate the transition from the maintenance of pluripotency to the commitment to certain lineages (Kalkan and Smith, 2014)?

In order to set up the TF-TF network, we have to identify the TF genes in mice. This was done using the Animal Transcription Factor Database (ATFDB) (Zhang et al., 2011). Out of 1458 genes predicted to be TFs by the ATFDB, 1297 could be mapped to the Gene St based microrray data set.

We obtained a literature-based network containing curated, high confidence interactions among TFs from the PluriNetWork (Som et al., 2010) that is aimed specifically at understanding gene regulation governing pluripotency and reprogramming. Apart from a further benchmark of the predicted networks this network is used to compare the topological properties of the predicted networks to the current knowledge about the real network. We removed loops from the literature network because they cannot be identified by an analysis based on co-expression.

Calculating the improvement of the AUPR for all three algorithms showed that partial correlation yields the lowest improvement while the predictions of CLR overlap the most with the literature network (Tab. 4.4). For the subsequent analysis of the network topologies we restricted the networks to the top 0.1% of interactions for each predicted network. To roughly estimate a lower bound on the precision of the networks obtained in this way we calculated the precision of the top 0.5% of interactions with respect to the literature network. We chose a cutoff of 0.5% because the overlap with the literature network consists of only 90 nodes implying that the top 0.5% correspond to just 40 interactions. At lower thresholds the calculated precision would fluctuate too strongly for slightly different predictions. At the 0.5% threshold we obtained precisions of 22% for partial correlation and 27% for CLR and ARACNE. We conclude that a lower threshold for the precision at the 0.1% level is given by 20 to 25%. It should however be noted that the majority of the TFs in the predicted networks are not present in the literature network so that the precision of a large part of the predicted network can not be determined.

The predicted networks are shown together with the literature network in Fig. 4.6. They obviously exhibit very different topologies. We next compared the topological properties of the literature network and the predicted networks by quantifying some standard measures (Tab. 4.5). The degree distribution (Fig.

Table 4.5: Quantification of topological properties of the predicted networks. The networks predicted by the algorithms CLR, partial correlation in the pls implementation and ARACNE with 0.50 as cutoff parameters are compared to the literature network. Mean rank OSN (Oct4, Sox2, Nanog) denotes the mean degree rank of the core triad, with low values indicating high degrees, % in largest component the fraction of vertices contained in the largest connected component, no connected components the number of connected components, modularity edge betweenness the modularity of the graph according to the edge betweenness community measure, modularity fastgreedy the modularity of the graph according to the fastgreedy community measure, diameter the graph diameter and transitivity the graph transitivity.

	clr	pcor.pls	aracne_50	literature
mean rank OSN	172.33	66	7	2
% in largest component	0.69	0.95	0.78	0.99
no connected components	61	12	22	2
modularity edge betweenness	0.72	0.56	0.21	0.12
modularity fastgreedy	0.72	0.54	0.29	0.34
diameter	22	10	9	5
transitivity	0.48	0.26	0.53	0.16
degree correlation	0.63	-0.01	0.12	-0.47

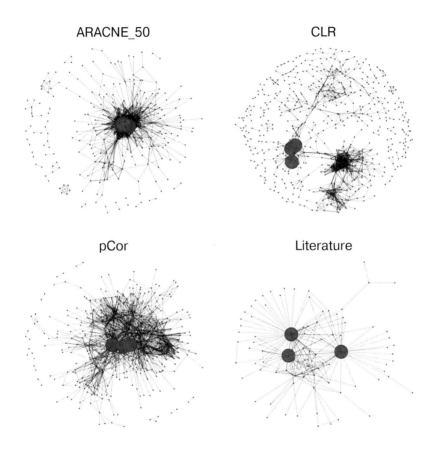

Figure 4.6: Comparison of the topologies of the predicted TF-TF networks with the literature-based network for the top ranking algorithms. The transcription factors Pou5f1, Sox2 and Nanog are indicated by magenta dots.

4.7) of the literature network shows an approximately linear decrease in the log-log-plot. Contrary to this the predicted networks show a marked deviation from a linear degree distribution for the highest degrees. The linear degree distribution of the literature network up to the highest degrees is reflected by the fact that the most important nodes, Nanog and Pou5f1 are involved in 50% of all interactions. Their targets typically have no interactions among themselves, leading to a star-like structure centered on Nanog and Pou5f1 (Fig. 4.6). This fact is reflected in the low transitivity of the literature network when compared with the predicted networks. A related quantity, the degree correlation coefficient indicates if nodes of high degree are typically connected to other nodes of high degree (positive degree correlation) or to nodes of low degree (negative degree correlation). The degree correlation coefficient of -0.47 confirms the star-like structure.

In the ARACNE network, interactions are focused on a small fraction of all nodes, which form the center of the largest connected component. The center of this connected component is formed by Oct4, Sox2 and Nanog among others. This is reflected in the very low mean rank of the degrees of this core triad. Similar to the literature network, the ARACNE network has a low modularity and also a relatively low diameter. It differs from the literature network in its degree correlation, which is positive. This fact points towards a hierarchy of nodes, with the highest degree nodes connected to the next highest degree nodes and so on.

The network predicted by partial correlation has the highest fraction of nodes with small degree among the predicted networks. Also, the degree correlation is near zero, indicating that there is little substructure in the largest connected component. There are no communities which are only connected to other communities via gateway nodes, resulting in a comparably low diameter and the largest connected component of all predicted networks. The mean rank of the degree of the core triad is higher than for ARACNE, reflecting a less central place of these TFs in the network.

Finally the network predicted by the CLR algorithm shows the most pronounced modular structure with groups of nodes that have high connectivity among themselves but low connectivity with outside nodes. The modular structure is reflected in the highest modularity index of all predicted networks as well as the highest transitivity and degree correlation. Though the core triad is located in a community of nodes with high connectivity, the smallness of this community leads to the highest mean rank of the degree of OSN among the predicted networks. The

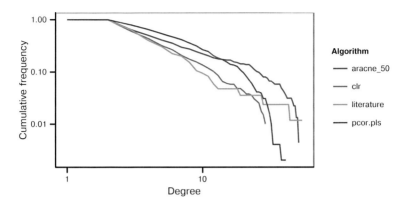

Figure 4.7: Degree distribution for the TF-TF gene regulatory networks predicted by the indicated algorithms, wit aracne$_{50}$ denoting the ARACNE algorithm with 0.5 as cutoff parameter and pcor.pls denoting partial correlation in the pls implementation.

large diameter of the CLR network is also a consequence of the modular structure. Nodes from different communities can only be connected by paths traversing the few gateway nodes that connect communities.

It is known that reconstruction algorithms in general tend to enrich different types of motifs (Marbach et al., 2012), affecting the topology of the predicted network. Some of the different topological features of the predicted networks can be explicitly traced back to the way the employed algorithm works. Networks induced by correlation are more transitive than random networks whereas those induced by partial correlation are less transitive than random networks (Zalesky et al., 2012). The ARACNE algorithm also directly influences transitivity by cutting all weakest links in triangles unless their strength is above the tolerance parameter. Here, ARACNE shows rather high transitivity because the cutoff parameter τ was set to the large value of 0.5. The CLR algorithm does not influence the topology in an explicit way, but induces a network that exhibits a unique modular structure among the predicted networks. In comparison the literature network is shaped strongly by the investigations into the role of Pou5f1 and Nanog and to a lesser degree Sox2. Thus its structure may to some degree also be a consequence of a publication bias. Taking into account the bias introduced by the different algorithms and the extreme focus of the literature network on the core triad, it seems impossible at the moment to draw definitive conclusions about the topology of the TF-TF network in mESC.

4.7 Discussion and outlook

Understanding how development is controlled at the molecular level depends to a large degree on understanding gene regulatory networks (Peter and Davidson, 2011). The switch in gene expression from the pluripotent state to the primed state in mouse ESC can serve as a model for a developmental process. With the ESC model we are in the fortunate situation of having access to a large body of published transcriptome data from which regulatory interactions can be deduced. Network reconstruction based on co-expression measures seems to be a promising approach for deducing interactions from this data.

Here, we analyzed the question how useful different reconstruction algorithms are for inferring the gene regulatory network of differentiating mouse ESC. In order to answer this question we determined the overlap of interactions predicted by different algorithms with interaction evidence based on high-throughput experiments. We could show that the algorithms that perform well with respect to evidence from ChIP-seq and transcription factor perturbation data all make use of pruning schemes that address the problem of indirect links. One of the top-performing algorithms, partial correlation, proved to be particularly interesting because its predictions are quite distinct from other well-performing algorithms. Finally, we analyzed the topology of the TF-TF regulatory network predicted by the three top-performing algorithms. We observed markedly different topologies whose features can be traced back to the design of the algorithm. Thus, while evidence from high-throughput data supports the usefulness of the predictions based on co-expression, the observed network structure may to some degree be an artifact of the employed algorithm. This is one of the aspects currently limiting the usefulness of network reconstruction.

Network reconstruction has been approached with differing scope and multiple tools. In particular, large scale networks reconstructed in this work and small scale networks in the literature are reconstructed with different goals in mind. Small scale networks are often used to integrate information about multiple perturbation experiments into a single model. This model can then be used to predict untested perturbations without tedious or costly experiments. This approach to molecular networks has proved to be fruitful in devising ways to manipulate cellular signal transduction (Klinger et al., 2013). In the field of stem cells, it can be used to understand how the wiring between signaling pathways and pluripotency-associated factors determines maintenance of the pluripotent state under different culture conditions (Dunn et al., 2014; Xu et al., 2014).

This binary decision to maintain pluripotency or start differentiation yields two clearly discernible phenotypes, with which a small scale network predicting phenotypes can be trained relatively easily. For the gene regulatory network controlling differentiation, defining distinct phenotypic outcomes is much harder. Thus the approach used for the network maintaining pluripotency can not be easily extended to the interactions between pluripotency-associated factors and lineage specific factors.

Large scale networks have often been developed with the goal of generating lists of candidate genes for follow-up experiments. This has also been one of the main applications in the stem cell literature (Cegli et al., 2013; Dowell et al., 2013). Another aspect that has been studied with large scale networks is the overall structure of the network and its relation to selective pressures that gave rise to the observed network structure (Davidson, 2010). Important concept in this context are overrepresented motifs or the frequency with which feedbacks are encountered in biological networks.

The current ways in which large scale networks are used and the limits in prediction quality observed in this work make evident that the goal of obtaining a regulatory network that can serve as a basis of a dynamical model of gene expression is still distant. How can progress in this direction be made? New data may significantly enhance our ability to infer networks. In particular, single cell data may open up the possibility to observe distinct states of a network which get blurred in bulk data (Buganim et al., 2012; Trott et al., 2012; Guo et al., 2010). This may help to resolve the temporal sequence of gene silencing and activation. Perturbations in conjunction with well resolved time series may also yield important insights since one can separate early, presumably direct, from late, presumably indirect effects. The currently available transcription factor perturbation data is usually optimized for detecting even weakly deregulated genes (Nishiyama et al., 2009). Since this is achieved by assessing the transcriptome days after the start of the perturbation, a large amount of differential expression is caused by indirect effects.

To allow for a more unbiased evaluation of network reconstruction efforts, a higher overlap between ChIP-seq-based transcription factor binding assays and transcription factor perturbation data would be very helpful. Intersecting genes whose promoters are bound by a transcription factor and that are differentially expressed when this transcription factor is perturbed would yield a high confidence set of targets.

The lack of overlapping gold standard data highlights the need for methodological work such as the one presented here. One of the major obstacles for benchmarking network reconstruction algorithms is the fact that there is still little knowledge which transcription factors are actually amenable to this approach (Margolin and Califano, 2007). Thus, an important step forward would be to determine the preconditions a transcription factor has to satisfy so that targets can in principle be inferred from co-expression with this transcription factor. With research too often focused on finding new regulatory mechanisms, such methodologically relevant work should receive more attention.

5 Conclusions

The transcriptome of differentiating mouse embryonic stem cells is shaped by multiple regulatory mechanisms. By analyzing patterns in the transcriptome, we can infer properties of these mechanisms. This will help to understand how cells achieve a robust, highly reproducible expression program.

An important aspect of gene regulation is fine-tuning of expression which ensures that interacting genes are expressed in appropriate ratios. In chapter 2, we studied to which extent expression on the X chromosome is upregulated to compensate for the monoallelic expression of X-linked genes. We showed that expression changes of X-linked genes depend on their evolutionary history. While acquired genes are not upregulated during differentiation with respect to the autosomes, ancestral genes exhibit upregulation irrespective of whether they are lowly or highly expressed. This indicates that at least during differentiation, the dosage of X-linked genes is increased to arrive at the same average expression levels as that of autosomal genes.

Whether this upregulation can be observed in somatic cells has been hotly debated in the literature (see Sec. 2.1.1). It is less controversial that there exist selection pressures for coordinated expression among certain groups of genes. For X-linked genes, it was shown that genes involved in large complexes together with autosomal genes are upregulated to arrive at an average X:A ratio of 1 (Pessia et al., 2012). The need for a precisely regulated stoichiometry of genes that are members of the same complex likely extends to some degree also to genes that are members of the same signaling pathway (Witzel et al., 2012). On the other hand, pathways also show evidence for robustness with respect to fluctuations in the expression of its components. Important examples for robustness to protein fluctuations come from the MAPK pathway (Fritsche-Guenther et al., 2011) and bacterial chemotaxis (Alon et al., 1999; Løvdok et al., 2009).

As our understanding of cellular pathways increases, it becomes clear that cells employ both strategies to fine-tune expression and to buffer expression fluctuations. These complementary strategies are also evidenced by the low, but non-negligible fraction of known haplo-insufficient genes in mice of around 5-10% (Huang et al., 2010).

As indicated above, certain processes in a cell require precisely tuned gene expression. In the differentiating cell, multiple processes that alter gene expression are acting at the same time, raising the question how these processes are coordinated. In the second part of chapter 2 we studied the coupling of X inactivation and differentiation. Using statistical analyses we showed that female ESC are delayed in differentiation with respect to XO/XY ESC. This delay was associated with a lack of *de novo* methylation in female cells. Experiments that were subsequently performed could show that the statistical association is also a causal one: two active X chromosomes stabilize the pluripotent state via MAPK signaling and DNA methylation. Together the statistical analyses and the experiments established that in addition to the regulation of X inactivation by the pluripotency-maintaining factors, there is also a feedback of X inactivation on these factors.

The developmental checkpoint of X chromosome silencing resembles the phenomenon of diapause in rodent embryogenesis. Embryos in diapause remain in an unimplanted state, preventing progress in development in lactating females (Lopes et al., 2004). Recovery from diapause is dependent on LIF signaling, since embryos lacking LIF receptors are unable to resume development (Nichols et al., 2001). These features of the diapause checkpoint exhibit a striking resemblance to the checkpoint of X inactivation. Both checkpoints prevent the exit from self-renewal of the pluripotent state and both are mediated by signaling. Indeed, some overlap of the mechanisms can not be excluded since there is some evidence that the existence of a diapause state is a prerequisite for the existence of naive ESC (Welling and Geijsen, 2013).

Regardless of whether both checkpoints are mediated by overlapping mechanisms, they point to the flexibility of the differentiation process. Embryonic stem cells thus exhibit seemingly contradictory properties: the orderly execution of a predefined differentiation program and responsiveness to internal or external cues. In the language of Waddington's famous epigenetic landscape (Waddington, 1957), we may interpret our results as follows. The groove in which a differentiating cell descends the slope of the epigenetic landscape is fixed, but the speed of descent may be modulated.

The mutual control of differentiation and X inactivation also points to the complexity of gene regulation. Multiple, coupled processes are thus altering gene expression levels at the same time. However, the complexity of gene regulation is believed to be somewhat limited by the modular organization of genes into functional groups (Hartwell et al., 1999). In chapter 3, we described a mechanism that induces co-regulation of adjacent genes and thus may give rise to such functional modules of genes. Chromatin in mammalian cells is arranged in space such that it forms domains of preferential contact, called TADs, at a scale of hundreds of kilobases. This contact induces co-expression of genes located in the same domain. We analyzed this effect at the X inactivation center where we observed that genes having an activatory or repressive effect on Xist fall into two distinct domains. On a genome-wide scale we could detect a significant increase in co-expression for genes in the same domain. Also, data from a protein-protein-interaction database indicates that proteins interact preferentially with proteins that are coded by genes from the same domain.

Since the initial discovery of TADs, there have been multiple independent observations of the co-expression-inducing effect of TADs on a quantitative level (Dily et al., 2014) or on the level of shared expression context (Symmons et al., 2014; Andrey et al., 2013). Additionally, they are closely related to previously characterized lamina associated domains, with changes in lamina association often happening in a TAD-wide manner. Because of these observations, TADs have been called "the fundamental modular unit of gene regulation and genome organization" (Denholtz et al., 2013) at the length scale of around 1 Mb. Clearly, the existence of modular units of gene regulation has important consequences for differentiation and development. This has been impressively demonstrated for the Hoxd genes involved in limb development. Two adjacent TADs control two separate waves of gene expression, with genes of the first controlling development of the proximal part of the limb and genes of the second controlling the development of the distal part (Andrey et al., 2013).

The concepts discussed so far, fine-tuning of expression, feedback regulation and modularity are also relevant for one of the central mechanisms of gene regulation in cells, transcriptional control by DNA-binding factors. To understand how these concepts shape the action of transcription factors, the topology of the gene regulatory network is essential. In chapter 4, we compared different algorithms that can be used to reconstruct the gene regulatory network in ESC from transcriptome data. The reconstructed networks were ranked by their overlap with biological evidence for functional transcription factor binding. Finally, we analyzed the topology of the reconstructed networks showing the highest overlap with experimental evidence. We showed that these topologies are highly dependent on the design of the reconstruction algorithm. This limitation of network reconstruction is already known from network reconstruction in simple organisms (Marbach et al., 2012).

Further advances in network reconstruction depend on additional or improved experimental data. New data can serve two purposes, improved benchmarking of network reconstruction and more reliable reconstruction itself. For improved benchmarking, the bottleneck is formed by the availability of ChIP-seq data. In particular, a larger number of ChIP-seq data sets could be used to define transcription factor targets based on the intersection of TF-DNA interactions and differential expression upon TF perturbation. This intersection is currently considered the gold standard of functional TF-gene interaction (Cusanovich et al., 2014). Additional temporal and spatial resolution of TF binding can also be expected from new assays. As sequencing costs continuously decline, it becomes possible to obtain a dynamic picture of DNA binding using seq-technologies. With the two techniques DNase-seq and ATAC-seq, it becomes feasible to monitor the temporal change of genome-wide protein occupancy patterns (Buenrostro et al., 2013).

For more reliable reconstruction itself, single-cell data may be most relevant. Single cell data is needed to properly resolve the sequence of cellular states during differentiation. In bulk data, intermediate steps of the transcriptome get blurred by asynchronous differentiation. However, currently single cell sequencing data can only be used to detect rather highly expressed genes.

In addition to characterizing each of the regulatory mechanisms we discussed, future research should also focus on the interactions between these mechanisms. One of the most important aspects regarding these interactions is the causal relationship between different layers of regulation. In particular, one of the open questions in differentiation is what the causal relationship between TF binding, epigenetic changes and spatial rearrangement of chromatin is. This information is needed to understand the precise temporal sequence of differentiation events. With respect to the causal relationship between TF binding and chromatin state, pioneer factors are now known to play an important role. Pioneer factors are special transcription factors that can access DNA in compact chromatin without the help of other co-factors (Chen and Dent, 2014).

In this work, we have used statistical analyses to discover connections between different regulatory layers in differentiating mouse embryonic stem cells. Each of these layers was characterized by high-throughput data such as microarray data for the transcriptome or Hi-C data to characterize chromatin conformation. From the point of view of systems biology, the greatest challenge lies in moving from statistical analyses to dynamical models. So far, large scale gene regulatory networks have resisted attempts to formulate meaningful dynamical models that can describe the temporal sequence of such complex events as cellular differentiation.

The challenge becomes even greater when multiple regulatory layers should be integrated into such a model. It remains to be seen whether the ever increasing amount of data will allow progress in this direction.

Bibliography

Akhtar, W., de Jong, J., Pindyurin, A. V., Pagie, L., Meuleman, W., de Ridder, J., Berns, A., Wessels, L. F. A., van Lohuizen, M., and van Steensel, B. (2013). Chromatin Position Effects Assayed by Thousands of Reporters Integrated in Parallel. *Cell*, 154(4):914–927.

Allen, J. D., Xie, Y., Chen, M., Girard, L., and Xiao, G. (2012). Comparing Statistical Methods for Constructing Large Scale Gene Networks. *PLoS ONE*, 7(1):e29348.

Allocco, D. J., Kohane, I. S., and Butte, A. J. (2004). Quantifying the relationship between co-expression, co-regulation and gene function. *BMC Bioinformatics*, 5(1):18.

Alon, U., Surette, M. G., Barkai, N., and Leibler, S. (1999). Robustness in bacterial chemotaxis. *Nature*, 397(6715):168–171.

Anchang, B., Sadeh, M. J., Jacob, J., Tresch, A., Vlad, M. O., Oefner, P. J., and Spang, R. (2009). Modeling the Temporal Interplay of Molecular Signaling and Gene Expression by Using Dynamic Nested Effects Models. *Proceedings of the National Academy of Sciences*, 106(16):6447–6452.

Andrey, G., Montavon, T., Mascrez, B., Gonzalez, F., Noordermeer, D., Leleu, M., Trono, D., Spitz, F., and Duboule, D. (2013). A Switch Between Topological Domains Underlies HoxD Genes Collinearity in Mouse Limbs. *Science*, 340(6137):1234167.

Augui, S., Nora, E. P., and Heard, E. (2011). Regulation of X-chromosome inactivation by the X-inactivation centre. *Nature Reviews Genetics*, 12(6):429–442.

Avilion, A. A., Nicolis, S. K., Pevny, L. H., Perez, L., Vivian, N., and Lovell-Badge, R. (2003). Multipotent cell lineages in early mouse development depend on SOX2 function. *Genes & Development*, 17(1):126–140.

Bansal, M., Belcastro, V., Ambesi-Impiombato, A., and di Bernardo, D. (2007). How to infer gene networks from expression profiles. *Molecular Systems Biology*, 3(1):n/a–n/a.

Basso, K., Margolin, A. A., Stolovitzky, G., Klein, U., Dalla-Favera, R., and Califano, A. (2005). Reverse engineering of regulatory networks in human B cells. *Nature Genetics*, 37(4):382–390.

Boroviak, T., Loos, R., Bertone, P., Smith, A., and Nichols, J. (2014). The ability of inner-cell-mass cells to self-renew as embryonic stem cells is acquired following epiblast specification. *Nature Cell Biology*, 16(6):513–525.

Boyd, K., Eng, K. H., and Page, C. D. (2013). Area under the Precision-Recall Curve: Point Estimates and Confidence Intervals. In Blockeel, H., Kersting, K., Nijssen, S., and Železný, F., editors, *Machine Learning and Knowledge Discovery in Databases*, number 8190 in Lecture Notes in Computer Science, pages 451–466. Springer Berlin Heidelberg.

Buecker, C., Srinivasan, R., Wu, Z., Calo, E., Acampora, D., Faial, T., Simeone, A., Tan, M., Swigut, T., and Wysocka, J. (2014). Reorganization of Enhancer Patterns in Transition from Naive to Primed Pluripotency. *Cell Stem Cell*, 14(6):838–853.

Buenrostro, J. D., Giresi, P. G., Zaba, L. C., Chang, H. Y., and Greenleaf, W. J. (2013). Transposition of native chromatin for fast and sensitive epigenomic profiling of open chromatin, DNA-binding proteins and nucleosome position. *Nature Methods*, 10(12):1213–1218.

Buganim, Y., Faddah, D. A., Cheng, A. W., Itskovich, E., Markoulaki, S., Ganz, K., Klemm, S. L., van Oudenaarden, A., and Jaenisch, R. (2012). Single-Cell Expression Analyses during Cellular Reprogramming Reveal an Early Stochastic and a Late Hierarchic Phase. *Cell*, 150(6):1209–1222.

Burgoyne, P. S., Thornhill, A. R., Boudrean, S. K., Darling, S. M., Bishop, C. E., Evans, E. P., Capel, B., and Mittwoch, U. (1995). The Genetic Basis of XX-XY Differences Present before Gonadal Sex Differentiation in the Mouse [and Discussion]. *Philosophical Transactions of the Royal Society of London B: Biological Sciences*, 350(1333):253–261.

Cahan, P., Li, H., Morris, S. A., Lummertz da Rocha, E., Daley, G. Q., and Collins, J. J. (2014). CellNet: Network Biology Applied to Stem Cell Engineering. *Cell*, 158(4):903–915.

Carter, A. C., Davis-Dusenbery, B. N., Koszka, K., Ichida, J. K., and Eggan, K. (2014). Nanog-Independent Reprogramming to iPSCs with Canonical Factors. *Stem Cell Reports*, 2(2):119–126.

Castagné, R., Rotival, M., Zeller, T., Wild, P. S., Truong, V., Trégouët, D.-A., Munzel, T., Ziegler, A., Cambien, F., Blankenberg, S., and Tiret, L. (2011). The Choice of the Filtering Method in Microarrays Affects the Inference Regarding Dosage Compensation of the Active X-Chromosome. *PLoS ONE*, 6(9):e23956.

Castelo, R. and Roverato, A. (2009). Reverse Engineering Molecular Regulatory Networks from Microarray Data with qp-Graphs. *Journal of Computational Biology*, 16(2):213–227.

Cegli, R. D., Iacobacci, S., Flore, G., Gambardella, G., Mao, L., Cutillo, L., Lauria, M., Klose, J., Illingworth, E., Banfi, S., and Bernardo, D. d. (2013). Reverse engineering a mouse embryonic stem cell-specific transcriptional network reveals a new modulator of neuronal differentiation. *Nucleic Acids Research*, 41(2):711–726.

Chen, T. and Dent, S. Y. R. (2014). Chromatin modifiers and remodellers: regulators of cellular differentiation. *Nature Reviews Genetics*, 15(2):93–106.

Ciofani, M., Madar, A., Galan, C., Sellars, M., Mace, K., Pauli, F., Agarwal, A., Huang, W., Parkurst, C. N., Muratet, M., Newberry, K. M., Meadows, S., Greenfield, A., Yang, Y., Jain, P., Kirigin, F. K., Birchmeier, C., Wagner, E. F., Murphy, K. M., Myers, R. M., Bonneau, R., and Littman, D. R. (2012). A Validated Regulatory Network for Th17 Cell Specification. *Cell*, 151(2):289–303.

Clerc, P. and Avner, P. (1998). Role of the region 3' to Xist exon 6 in the counting process of X-chromosome inactivation. *Nature Genetics*, 19(3):249–253.

Conti, L., Pollard, S. M., Gorba, T., Reitano, E., Toselli, M., Biella, G., Sun, Y., Sanzone, S., Ying, Q.-L., Cattaneo, E., and Smith, A. (2005). Niche-Independent Symmetrical Self-Renewal of a Mammalian Tissue Stem Cell. *PLoS Biol*, 3(9):e283.

Correa-Cerro, L. S., Piao, Y., Sharov, A. A., Nishiyama, A., Cadet, J. S., Yu, H., Sharova, L. V., Xin, L., Hoang, H. G., Thomas, M., Qian, Y., Dudekula, D. B., Meyers, E., Binder, B. Y., Mowrer, G., Bassey, U., Longo, D. L., Schlessinger, D., and Ko, M. S. H. (2011). Generation of mouse ES cell lines engineered for the forced induction of transcription factors. *Scientific Reports*, 1.

Cusanovich, D. A., Pavlovic, B., Pritchard, J. K., and Gilad, Y. (2014). The Functional Consequences of Variation in Transcription Factor Binding. *PLoS Genet*, 10(3):e1004226.

Davidson, E. H. (2010). Emerging properties of animal gene regulatory networks. *Nature*, 468(7326):911–920.

Davis, J. and Goadrich, M. (2006). The Relationship Between Precision-Recall and ROC Curves. In *Proceedings of the 23rd International Conference on Machine Learning*, ICML '06, pages 233–240, New York, NY, USA. ACM.

de Jong, H. (2002). Modeling and Simulation of Genetic Regulatory Systems: A Literature Review. *Journal of Computational Biology*, 9(1):67–103.

Deng, X., Berletch, J. B., Ma, W., Nguyen, D. K., Hiatt, J. B., Noble, W. S., Shendure, J., and Disteche, C. M. (2013). Mammalian X Upregulation Is Associated with Enhanced Transcription Initiation, RNA Half-Life, and MOF-Mediated H4k16 Acetylation. *Developmental Cell*, 25(1):55–68.

Deng, X., Hiatt, J. B., Nguyen, D. K., Ercan, S., Sturgill, D., Hillier, L. W., Schlesinger, F., Davis, C. A., Reinke, V. J., Gingeras, T. R., Shendure, J., Waterston, R. H., Oliver, B., Lieb, J. D., and Disteche, C. M. (2011). Evidence for compensatory upregulation of expressed X-linked genes in mammals, Caenorhabditis elegans and Drosophila melanogaster. *Nature Genetics*, 43(12):1179–1185.

Denholtz, M., Bonora, G., Chronis, C., Splinter, E., de Laat, W., Ernst, J., Pellegrini, M., and Plath, K. (2013). Long-Range Chromatin Contacts in Embryonic Stem Cells Reveal a Role for Pluripotency Factors and Polycomb Proteins in Genome Organization. *Cell Stem Cell*, 13(5):602–616.

Deuve, J. L. and Avner, P. (2011). The Coupling of X-Chromosome Inactivation to Pluripotency. *Annual Review of Cell and Developmental Biology*, 27(1):611–629.

D'Hulst, C., Parvanova, I., Tomoiaga, D., Sapar, M. L., and Feinstein, P. (2013). Fast Quantitative Real-Time PCR-Based Screening for Common Chromosomal Aneuploidies in Mouse Embryonic Stem Cells. *Stem Cell Reports*, 1(4):350–359.

Dily, F. L., Baù, D., Pohl, A., Vicent, G. P., Serra, F., Soronellas, D., Castellano, G., Wright, R. H. G., Ballare, C., Filion, G., Marti-Renom, M. A., and Beato, M. (2014). Distinct structural transitions of chromatin topological domains correlate with coordinated hormone-induced gene regulation. *Genes & Development*, 28(19):2151–2162.

Dixon, J. R., Selvaraj, S., Yue, F., Kim, A., Li, Y., Shen, Y., Hu, M., Liu, J. S., and Ren, B. (2012). Topological domains in mammalian genomes identified by analysis of chromatin interactions. *Nature*, 485(7398):376–380.

Dostie, J., Richmond, T. A., Arnaout, R. A., Selzer, R. R., Lee, W. L., Honan, T. A., Rubio, E. D., Krumm, A., Lamb, J., Nusbaum, C., Green, R. D., and Dekker, J. (2006). Chromosome Conformation Capture Carbon Copy (5c): A massively parallel solution for mapping interactions between genomic elements. *Genome Research*, 16(10):1299–1309.

Dowell, K. G., Simons, A. K., Wang, Z. Z., Yun, K., and Hibbs, M. A. (2013). Cell-Type-Specific Predictive Network Yields Novel Insights into Mouse Embryonic Stem Cell Self-Renewal and Cell Fate. *PLoS ONE*, 8(2):e56810.

Dunn, S.-J., Martello, G., Yordanov, B., Emmott, S., and Smith, A. G. (2014). Defining an essential transcription factor program for naïve pluripotency. *Science*, 344(6188):1156–1160.

Dupont, C. and Gribnau, J. (2013). Different flavors of X-chromosome inactivation in mammals. *Current Opinion in Cell Biology*, 25(3):314–321.

Eckner, A. (2012). Algorithms for Unevenly-Spaced Time Series: Moving Averages and Other Rolling Operators. http://www.eckner.com/papers/ts_alg.pdf.

Faith, J. J., Hayete, B., Thaden, J. T., Mogno, I., Wierzbowski, J., Cottarel, G., Kasif, S., Collins, J. J., and Gardner, T. S. (2007). Large-Scale Mapping and Validation of Escherichia coli Transcriptional Regulation from a Compendium of Expression Profiles. *PLoS Biol*, 5(1):e8.

Faunes, F., Hayward, P., Descalzo, S. M., Chatterjee, S. S., Balayo, T., Trott, J., Christoforou, A., Ferrer-Vaquer, A., Hadjantonakis, A.-K., Dasgupta, R., and Arias, A. M. (2013). A membrane-associated β-catenin/Oct4 complex correlates with ground-state pluripotency in mouse embryonic stem cells. *Development*, 140(6):1171–1183.

Filippova, D., Patro, R., Duggal, G., and Kingsford, C. (2014). Identification of alternative topological domains in chromatin. *Algorithms for Molecular Biology*, 9(1):14.

Fouse, S. D., Shen, Y., Pellegrini, M., Cole, S., Meissner, A., Neste, L. V., Jaenisch, R., and Fan, G. (2008). Promoter CpG Methylation Contributes to ES Cell Gene

Regulation in Parallel with Oct4/Nanog, PcG Complex, and Histone H3 K4/K27 Trimethylation. *Cell Stem Cell*, 2(2):160–169.

Franceschini, A., Szklarczyk, D., Frankild, S., Kuhn, M., Simonovic, M., Roth, A., Lin, J., Minguez, P., Bork, P., Mering, C. v., and Jensen, L. J. (2013). STRING v9.1: protein-protein interaction networks, with increased coverage and integration. *Nucleic Acids Research*, 41(D1):D808–D815.

Fritsche-Guenther, R., Witzel, F., Sieber, A., Herr, R., Schmidt, N., Braun, S., Brummer, T., Sers, C., and Blüthgen, N. (2011). Strong negative feedback from Erk to Raf confers robustness to MAPK signalling. *Molecular Systems Biology*, 7(1).

Fukuoka, Y., Inaoka, H., and Kohane, I. S. (2004). Inter-species differences of co-expression of neighboring genes in eukaryotic genomes. *BMC Genomics*, 5(1):4.

Gendrel, A.-V., Attia, M., Chen, C.-J., Diabangouaya, P., Servant, N., Barillot, E., and Heard, E. (2014). Developmental Dynamics and Disease Potential of Random Monoallelic Gene Expression. *Developmental Cell*, 28(4):366–380.

Gherman, A., Wang, R., and Avramopoulos, D. (2009). Orientation, distance, regulation and function of neighbouring genes. *Human Genomics*, 3(2):143.

Glass, L. and Kauffman, S. A. (1973). The logical analysis of continuous, non-linear biochemical control networks. *Journal of Theoretical Biology*, 39(1):103–129.

Göke, J., Jung, M., Behrens, S., Chavez, L., O'Keeffe, S., Timmermann, B., Lehrach, H., Adjaye, J., and Vingron, M. (2011). Combinatorial Binding in Human and Mouse Embryonic Stem Cells Identifies Conserved Enhancers Active in Early Embryonic Development. *PLoS Computational Biology*, 7(12):e1002304.

Gómez-Díaz, E. and Corces, V. G. (2014). Architectural proteins: regulators of 3d genome organization in cell fate. *Trends in Cell Biology*, 24(11):703–711.

Gontan, C., Achame, E. M., Demmers, J., Barakat, T. S., Rentmeester, E., van IJcken, W., Grootegoed, J. A., and Gribnau, J. (2012). RNF12 initiates X-chromosome inactivation by targeting REX1 for degradation. *Nature*, 485(7398):386–390.

Graf, T. and Stadtfeld, M. (2008). Heterogeneity of Embryonic and Adult Stem Cells. *Cell Stem Cell*, 3(5):480–483.

Guo, G., Huss, M., Tong, G. Q., Wang, C., Li Sun, L., Clarke, N. D., and Robson, P. (2010). Resolution of Cell Fate Decisions Revealed by Single-Cell Gene Expression Analysis from Zygote to Blastocyst. *Developmental Cell*, 18(4):675–685.

Habibi, E., Brinkman, A. B., Arand, J., Kroeze, L. I., Kerstens, H. H. D., Matarese, F., Lepikhov, K., Gut, M., Brun-Heath, I., Hubner, N. C., Benedetti, R., Altucci, L., Jansen, J. H., Walter, J., Gut, I. G., Marks, H., and Stunnenberg, H. G. (2013). Whole-Genome Bisulfite Sequencing of Two Distinct Interconvertible DNA Methylomes of Mouse Embryonic Stem Cells. *Cell Stem Cell*, 13(3):360–369.

Hackett, J. A. and Surani, M. A. (2014). Regulatory Principles of Pluripotency: From the Ground State Up. *Cell Stem Cell*, 15(4):416–430.

Hartwell, L. H., Hopfield, J. J., Leibler, S., and Murray, A. W. (1999). From molecular to modular cell biology. *Nature*, 402:C47–C52.

Hayashi, K., Ohta, H., Kurimoto, K., Aramaki, S., and Saitou, M. (2011). Reconstitution of the Mouse Germ Cell Specification Pathway in Culture by Pluripotent Stem Cells. *Cell*, 146(4):519–532.

Haynes, B. C., Maier, E. J., Kramer, M. H., Wang, P. I., Brown, H., and Brent, M. R. (2013). Mapping functional transcription factor networks from gene expression data. *Genome Research*, 23(8):1319–1328.

He, F., Balling, R., and Zeng, A.-P. (2009). Reverse engineering and verification of gene networks: Principles, assumptions, and limitations of present methods and future perspectives. *Journal of Biotechnology*, 144(3):190–203.

He, X., Chen, X., Xiong, Y., Chen, Z., Wang, X., Shi, S., Wang, X., and Zhang, J. (2011). He et al. reply. *Nature Genetics*, 43(12):1171–1172.

Heard, E., Mongelard, F., Arnaud, D., and Avner, P. (1999). Xist Yeast Artificial Chromosome Transgenes Function as X-Inactivation Centers Only in Multicopy Arrays and Not as Single Copies. *Molecular and Cellular Biology*, 19(4):3156–3166.

Herzing, L. B. K., Romer, J. T., Horn, J. M., and Ashworth, A. (1997). Xist has properties of the X-chromosome inactivation centre. *Nature*, 386(6622):272–275.

Honkela, A., Girardot, C., Gustafson, E. H., Liu, Y.-H., Furlong, E. E. M., Lawrence, N. D., and Rattray, M. (2010). Model-based method for transcription factor target identification with limited data. *Proceedings of the National Academy of Sciences*, 107(17):7793–7798.

Huang, G., Ye, S., Zhou, X., Liu, D., and Ying, Q.-L. (2015). Molecular basis of embryonic stem cell self-renewal: from signaling pathways to pluripotency network. *Cellular and Molecular Life Sciences*, pages 1–17.

Huang, N., Lee, I., Marcotte, E. M., and Hurles, M. E. (2010). Characterising and Predicting Haploinsufficiency in the Human Genome. *PLoS Genet*, 6(10):e1001154.

Huerta, A. M., Salgado, H., Thieffry, D., and Collado-Vides, J. (1998). RegulonDB: A database on transcriptional regulation in Escherichia coli. *Nucleic Acids Research*, 26(1):55–59.

Hurst, L. D., Pál, C., and Lercher, M. J. (2004). The evolutionary dynamics of eukaryotic gene order. *Nature Reviews Genetics*, 5(4):299–310.

Jang, I. S., Margolin, A., and Califano, A. (2013). hARACNe: improving the accuracy of regulatory model reverse engineering via higher-order data processing inequality tests. *Interface Focus*, 3(4):20130011.

Joffe, B., Leonhardt, H., and Solovei, I. (2010). Differentiation and large scale spatial organization of the genome. *Current Opinion in Genetics & Development*, 20(5):562–569.

Joshi, A., Beck, Y., and Michoel, T. (2014). Multi-species network inference improves gene regulatory network reconstruction for early embryonic development in Drosophila. *arXiv:1407.6554 [q-bio]*. arXiv: 1407.6554.

Jue, N. K., Murphy, M. B., Kasowitz, S. D., Qureshi, S. M., Obergfell, C. J., Elsisi, S., Foley, R. J., O'Neill, R. J., and O'Neill, M. J. (2013). Determination of dosage compensation of the mammalian X chromosome by RNA-seq is dependent on analytical approach. *BMC Genomics*, 14(1):150.

Julien, P., Brawand, D., Soumillon, M., Necsulea, A., Liechti, A., Schütz, F., Daish, T., Grützner, F., and Kaessmann, H. (2012). Mechanisms and Evolutionary Patterns of Mammalian and Avian Dosage Compensation. *PLoS Biol*, 10(5):e1001328.

Kaderali, L. and Radde, N. (2008). Inferring Gene Regulatory Networks from Expression Data. In Kelemen, A., Abraham, A., and Chen, Y., editors, *Computational Intelligence in Bioinformatics*, number 94 in Studies in Computational Intelligence, pages 33–74. Springer Berlin Heidelberg.

Kalkan, T. and Smith, A. (2014). Mapping the route from naive pluripotency to lineage specification. *Philosophical Transactions of the Royal Society of London B: Biological Sciences*, 369(1657):20130540.

Karimi, M. M., Goyal, P., Maksakova, I. A., Bilenky, M., Leung, D., Tang, J. X., Shinkai, Y., Mager, D. L., Jones, S., Hirst, M., and Lorincz, M. C. (2011). DNA Methylation and SETDB1/H3k9me3 Regulate Predominantly Distinct Sets of Genes, Retroelements, and Chimeric Transcripts in mESCs. *Cell Stem Cell*, 8(6):676–687.

Kharchenko, P. V., Xi, R., and Park, P. J. (2011). Evidence for dosage compensation between the X chromosome and autosomes in mammals. *Nature Genetics*, 43(12):1167–1169.

Kim, J., Chu, J., Shen, X., Wang, J., and Orkin, S. H. (2008). An Extended Transcriptional Network for Pluripotency of Embryonic Stem Cells. *Cell*, 132(6):1049–1061.

Kim, Y. M., Lee, J.-Y., Xia, L., Mulvihill, J. J., and Li, S. (2013). Trisomy 8: a common finding in mouse embryonic stem (ES) cell lines. *Molecular Cytogenetics*, 6(1):3.

Klinger, B., Sieber, A., Fritsche-Guenther, R., Witzel, F., Berry, L., Schumacher, D., Yan, Y., Durek, P., Merchant, M., Schäfer, R., Sers, C., and Blüthgen, N. (2013). Network quantification of EGFR signaling unveils potential for targeted combination therapy. *Molecular Systems Biology*, 9(1):673.

Krämer, N., Schäfer, J., and Boulesteix, A.-L. (2009). Regularized estimation of large-scale gene association networks using graphical Gaussian models. *BMC Bioinformatics*, 10(1):384.

Lee, J. M. and Sonnhammer, E. L. L. (2003). Genomic Gene Clustering Analysis of Pathways in Eukaryotes. *Genome Research*, 13(5):875–882.

Lee, J. T. (2011). Gracefully ageing at 50, X-chromosome inactivation becomes a paradigm for RNA and chromatin control. *Nature Reviews Molecular Cell Biology*, 12(12):815–826.

Leeb, M., Dietmann, S., Paramor, M., Niwa, H., and Smith, A. (2014). Genetic Exploration of the Exit from Self-Renewal Using Haploid Embryonic Stem Cells. *Cell Stem Cell*, 14(3):385–393.

Leitch, H. G., McEwen, K. R., Turp, A., Encheva, V., Carroll, T., Grabole, N., Mansfield, W., Nashun, B., Knezovich, J. G., Smith, A., Surani, M. A., and Hajkova, P. (2013). Naive pluripotency is associated with global DNA hypomethylation. *Nature Structural & Molecular Biology*, 20(3):311–316.

Li, J., Wei, H., Liu, T., and Zhao, P. X. (2013). GPLEXUS: enabling genome-scale gene association network reconstruction and analysis for very large-scale expression data. *Nucleic Acids Research*, page gkt983.

Li, Y.-Y., Yu, H., Guo, Z.-M., Guo, T.-Q., Tu, K., and Li, Y.-X. (2006). Systematic Analysis of Head-to-Head Gene Organization: Evolutionary Conservation and Potential Biological Relevance. *PLoS Comput Biol*, 2(7):e74.

Liang, X.-J., Xia, Z., Zhang, L.-W., and Wu, F.-X. (2012). Inference of gene regulatory subnetworks from time course gene expression data. *BMC Bioinformatics*, 13(Suppl 9):S3.

Lieberman-Aiden, E., Berkum, N. L. v., Williams, L., Imakaev, M., Ragoczy, T., Telling, A., Amit, I., Lajoie, B. R., Sabo, P. J., Dorschner, M. O., Sandstrom, R., Bernstein, B., Bender, M. A., Groudine, M., Gnirke, A., Stamatoyannopoulos, J., Mirny, L. A., Lander, E. S., and Dekker, J. (2009). Comprehensive Mapping of Long-Range Interactions Reveals Folding Principles of the Human Genome. *Science*, 326(5950):289–293.

Lin, F., Xing, K., Zhang, J., and He, X. (2012). Expression reduction in mammalian X chromosome evolution refutes Ohno's hypothesis of dosage compensation. *Proceedings of the National Academy of Sciences*, 109(29):11752–11757.

Lin, H., Gupta, V., VerMilyea, M. D., Falciani, F., Lee, J. T., O'Neill, L. P., and Turner, B. M. (2007). Dosage Compensation in the Mouse Balances Up-Regulation and Silencing of X-Linked Genes. *PLoS Biol*, 5(12):e326.

Lin, H., Halsall, J. A., Antczak, P., O'Neill, L. P., Falciani, F., and Turner, B. M. (2011). Relative overexpression of X-linked genes in mouse embryonic stem cells is consistent with Ohno's hypothesis. *Nature Genetics*, 43(12):1169–1170.

Loh, Y.-H., Wu, Q., Chew, J.-L., Vega, V. B., Zhang, W., Chen, X., Bourque, G., George, J., Leong, B., Liu, J., Wong, K.-Y., Sung, K. W., Lee, C. W. H.,

Zhao, X.-D., Chiu, K.-P., Lipovich, L., Kuznetsov, V. A., Robson, P., Stanton, L. W., Wei, C.-L., Ruan, Y., Lim, B., and Ng, H.-H. (2006). The Oct4 and Nanog transcription network regulates pluripotency in mouse embryonic stem cells. *Nature Genetics*, 38(4):431–440.

Lopes, F. L., Desmarais, J. A., and Murphy, B. D. (2004). Embryonic diapause and its regulation. *Reproduction (Cambridge, England)*, 128(6):669–678.

Løvdok, L., Bentele, K., Vladimirov, N., Müller, A., Pop, F. S., Lebiedz, D., Kollmann, M., and Sourjik, V. (2009). Role of translational coupling in robustness of bacterial chemotaxis pathway. *PLoS biology*, 7(8):e1000171.

Mank, J. E., Hosken, D. J., and Wedell, N. (2011). Some Inconvenient Truths About Sex Chromosome Dosage Compensation and the Potential Role of Sexual Conflict. *Evolution*, 65(8):2133–2144.

Marbach, D., Costello, J. C., Küffner, R., Vega, N. M., Prill, R. J., Camacho, D. M., Allison, K. R., The DREAM5 Consortium, Kellis, M., Collins, J. J., and Stolovitzky, G. (2012). Wisdom of crowds for robust gene network inference. *Nature Methods*, 9(8):796–804.

Marbach, D., Prill, R. J., Schaffter, T., Mattiussi, C., Floreano, D., and Stolovitzky, G. (2010). Revealing strengths and weaknesses of methods for gene network inference. *Proceedings of the National Academy of Sciences*, 107(14):6286–6291.

Margolin, A. A. and Califano, A. (2007). Theory and Limitations of Genetic Network Inference from Microarray Data. *Annals of the New York Academy of Sciences*, 1115(1):51–72.

Margolin, A. A., Nemenman, I., Basso, K., Wiggins, C., Stolovitzky, G., Favera, R. D., and Califano, A. (2006). ARACNE: An Algorithm for the Reconstruction of Gene Regulatory Networks in a Mammalian Cellular Context. *BMC Bioinformatics*, 7(Suppl 1):1–15.

Marks, H., Kalkan, T., Menafra, R., Denissov, S., Jones, K., Hofemeister, H., Nichols, J., Kranz, A., Francis Stewart, A., Smith, A., and Stunnenberg, H. G. (2012). The Transcriptional and Epigenomic Foundations of Ground State Pluripotency. *Cell*, 149(3):590–604.

Martello, G. and Smith, A. (2014). The Nature of Embryonic Stem Cells. *Annual Review of Cell and Developmental Biology*, 30(1):647–675.

Maston, G. A., Evans, S. K., and Green, M. R. (2006). Transcriptional Regulatory Elements in the Human Genome. *Annual Review of Genomics and Human Genetics*, 7(1):29–59.

Mering, C. v., Jensen, L. J., Snel, B., Hooper, S. D., Krupp, M., Foglierini, M., Jouffre, N., Huynen, M. A., and Bork, P. (2005). STRING: known and predicted protein–protein associations, integrated and transferred across organisms. *Nucleic Acids Research*, 33(suppl 1):D433–D437.

Meyer, P. E., Kontos, K., Lafitte, F., and Bontempi, G. (2007). Information-Theoretic Inference of Large Transcriptional Regulatory Networks. *EURASIP Journal on Bioinformatics and Systems Biology*, 2007(1):79879.

Michalak, P. (2008). Coexpression, coregulation, and cofunctionality of neighboring genes in eukaryotic genomes. *Genomics*, 91(3):243–248.

Milot, E., Strouboulis, J., Trimborn, T., Wijgerde, M., de Boer, E., Langeveld, A., Tan-Un, K., Vergeer, W., Yannoutsos, N., Grosveld, F., and Fraser, P. (1996). Heterochromatin Effects on the Frequency and Duration of LCR-Mediated Gene Transcription. *Cell*, 87(1):105–114.

Minkovsky, A., Patel, S., and Plath, K. (2012). Concise Review: Pluripotency and the Transcriptional Inactivation of the Female Mammalian X Chromosome. *STEM CELLS*, 30(1):48–54.

Morey, C., Kress, C., and Bickmore, W. A. (2009). Lack of bystander activation shows that localization exterior to chromosome territories is not sufficient to up-regulate gene expression. *Genome Research*, 19(7):1184–1194.

Nakaki, F. and Saitou, M. (2014). PRDM14: a unique regulator for pluripotency and epigenetic reprogramming. *Trends in Biochemical Sciences*, 39(6):289–298.

Navarro, P. and Avner, P. (2010). An embryonic story: Analysis of the gene regulative network controlling Xist expression in mouse embryonic stem cells. *BioEssays*, 32(7):581–588.

Navarro, P., Chambers, I., Karwacki-Neisius, V., Chureau, C., Morey, C., Rougeulle, C., and Avner, P. (2008). Molecular Coupling of Xist Regulation and Pluripotency. *Science*, 321(5896):1693–1695.

Navarro, P., Chantalat, S., Foglio, M., Chureau, C., Vigneau, S., Clerc, P., Avner, P., and Rougeulle, C. (2009). A role for non-coding Tsix transcription in partitioning chromatin domains within the mouse X-inactivation centre. *Epigenetics & Chromatin*, 2(1):8.

Navarro, P., Moffat, M., Mullin, N. P., and Chambers, I. (2011). The X-inactivation trans-activator Rnf12 is negatively regulated by pluripotency factors in embryonic stem cells. *Human Genetics*, 130(2):255–264.

Navarro, P., Page, D. R., Avner, P., and Rougeulle, C. (2006). Tsix-mediated epigenetic switch of a CTCF-flanked region of the Xist promoter determines the Xist transcription program. *Genes & Development*, 20(20):2787–2792.

Navarro, P., Pichard, S., Ciaudo, C., Avner, P., and Rougeulle, C. (2005). Tsix transcription across the Xist gene alters chromatin conformation without affecting Xist transcription: implications for X-chromosome inactivation. *Genes & Development*, 19(12):1474–1484.

Ng, H.-H. and Surani, M. A. (2011). The transcriptional and signalling networks of pluripotency. *Nature Cell Biology*, 13(5):490–496.

Nguyen, D. K. and Disteche, C. M. (2006). Dosage compensation of the active X chromosome in mammals. *Nature Genetics*, 38(1):47–53.

Nichols, J., Chambers, I., Taga, T., and Smith, A. (2001). Physiological rationale for responsiveness of mouse embryonic stem cells to gp130 cytokines. *Development*, 128(12):2333–2339.

Nichols, J. and Smith, A. (2009). Naive and Primed Pluripotent States. *Cell Stem Cell*, 4(6):487–492.

Nichols, J., Zevnik, B., Anastassiadis, K., Niwa, H., Klewe-Nebenius, D., Chambers, I., Schöler, H., and Smith, A. (1998). Formation of Pluripotent Stem Cells in the Mammalian Embryo Depends on the POU Transcription Factor Oct4. *Cell*, 95(3):379–391.

Nishiyama, A., Sharov, A. A., Piao, Y., Amano, M., Amano, T., Hoang, H. G., Binder, B. Y., Tapnio, R., Bassey, U., Malinou, J. N., Correa-Cerro, L. S., Yu, H., Xin, L., Meyers, E., Zalzman, M., Nakatake, Y., Stagg, C., Sharova, L., Qian, Y., Dudekula, D., Sheer, S., Cadet, J. S., Hirata, T., Yang, H.-T., Goldberg, I., Evans, M. K., Longo, D. L., Schlessinger, D., and Ko, M. S. H.

(2013). Systematic repression of transcription factors reveals limited patterns of gene expression changes in ES cells. *Scientific Reports*, 3.

Nishiyama, A., Xin, L., Sharov, A. A., Thomas, M., Mowrer, G., Meyers, E., Piao, Y., Mehta, S., Yee, S., Nakatake, Y., Stagg, C., Sharova, L., Correa-Cerro, L. S., Bassey, U., Hoang, H., Kim, E., Tapnio, R., Qian, Y., Dudekula, D., Zalzman, M., Li, M., Falco, G., Yang, H.-T., Lee, S.-L., Monti, M., Stanghellini, I., Islam, M. N., Nagaraja, R., Goldberg, I., Wang, W., Longo, D. L., Schlessinger, D., and Ko, M. S. H. (2009). Uncovering Early Response of Gene Regulatory Networks in ESCs by Systematic Induction of Transcription Factors. *Cell Stem Cell*, 5(4):420–433.

Nora, E. P., Dekker, J., and Heard, E. (2013). Segmental folding of chromosomes: A basis for structural and regulatory chromosomal neighborhoods? *BioEssays*, 35(9):818–828.

Nora, E. P., Lajoie, B. R., Schulz, E. G., Giorgetti, L., Okamoto, I., Servant, N., Piolot, T., Berkum, N. L. v., Meisig, J., Sedat, J., Gribnau, J., Barillot, E., Blüthgen, N., Dekker, J., and Heard, E. (2012). Spatial partitioning of the regulatory landscape of the X-inactivation centre. *Nature*, 485(7398):381–385.

Ohno, S. (1967). *Sex chromosomes and sex-linked genes*. Springer-Verlag.

Ong, C.-T. and Corces, V. G. (2014). CTCF: an architectural protein bridging genome topology and function. *Nature Reviews Genetics*, 15(4):234–246.

Ooi, S. K., Wolf, D., Hartung, O., Agarwal, S., Daley, G. Q., Goff, S. P., and Bestor, T. H. (2010). Dynamic instability of genomic methylation patterns in pluripotent stem cells. *Epigenetics & Chromatin*, 3(1):17.

Ouyang, Z., Zhou, Q., and Wong, W. H. (2009). ChIP-Seq of transcription factors predicts absolute and differential gene expression in embryonic stem cells. *Proceedings of the National Academy of Sciences*, 106(51):21521 –21526.

Peric-Hupkes, D., Meuleman, W., Pagie, L., Bruggeman, S. W. M., Solovei, I., Brugman, W., Gräf, S., Flicek, P., Kerkhoven, R. M., van Lohuizen, M., Reinders, M., Wessels, L., and van Steensel, B. (2010). Molecular Maps of the Reorganization of Genome-Nuclear Lamina Interactions during Differentiation. *Molecular Cell*, 38(4):603–613.

Perkins, J. R., Antunes-Martins, A., Calvo, M., Grist, J., Rust, W., Schmid, R., Hildebrandt, T., Kohl, M., Orengo, C., McMahon, S. B., and Bennett, D. L.

(2014). A comparison of RNA-seq and exon arrays for whole genome transcription profiling of the L5 spinal nerve transection model of neuropathic pain in the rat. *Molecular Pain*, 10(1):7.

Pessia, E., Engelstädter, J., and Marais, G. A. B. (2014). The evolution of X chromosome inactivation in mammals: the demise of Ohno's hypothesis? *Cellular and Molecular Life Sciences*, 71(8):1383–1394.

Pessia, E., Makino, T., Bailly-Bechet, M., McLysaght, A., and Marais, G. A. B. (2012). Mammalian X chromosome inactivation evolved as a dosage-compensation mechanism for dosage-sensitive genes on the X chromosome. *Proceedings of the National Academy of Sciences*, 109(14):5346–5351.

Peter, I. S. and Davidson, E. H. (2011). Evolution of Gene Regulatory Networks Controlling Body Plan Development. *Cell*, 144(6):970–985.

Pope, B. D., Ryba, T., Dileep, V., Yue, F., Wu, W., Denas, O., Vera, D. L., Wang, Y., Hansen, R. S., Canfield, T. K., Thurman, R. E., Cheng, Y., Gülsoy, G., Dennis, J. H., Snyder, M. P., Stamatoyannopoulos, J. A., Taylor, J., Hardison, R. C., Kahveci, T., Ren, B., and Gilbert, D. M. (2014). Topologically associating domains are stable units of replication-timing regulation. *Nature*, 515(7527):402–405.

Price, M. N., Huang, K. H., Arkin, A. P., and Alm, E. J. (2005). Operon formation is driven by co-regulation and not by horizontal gene transfer. *Genome Research*, 15(6):809–819.

Rastan, S. (1983). Non-random X-chromosome inactivation in mouse X-autosome translocation embryos—location of the inactivation centre. *Journal of Embryology and Experimental Morphology*, 78(1):1–22.

Reik, W. (2007). Stability and flexibility of epigenetic gene regulation in mammalian development. *Nature*, 447(7143):425–432.

Roy, S., Lagree, S., Hou, Z., Thomson, J. A., Stewart, R., and Gasch, A. P. (2013). Integrated Module and Gene-Specific Regulatory Inference Implicates Upstream Signaling Networks. *PLoS Comput Biol*, 9(10):e1003252.

Sado, T., Hoki, Y., and Sasaki, H. (2005). Tsix Silences Xist through Modification of Chromatin Structure. *Developmental Cell*, 9(1):159–165.

Sakaue, M., Ohta, H., Kumaki, Y., Oda, M., Sakaide, Y., Matsuoka, C., Yamagiwa, A., Niwa, H., Wakayama, T., and Okano, M. (2010). DNA Methylation Is Dispensable for the Growth and Survival of the Extraembryonic Lineages. *Current Biology*, 20(16):1452–1457.

Sales, G. and Romualdi, C. (2011). parmigene—a parallel R package for mutual information estimation and gene network reconstruction. *Bioinformatics*, 27(13):1876–1877.

Salgado, H., Gama-Castro, S., Peralta-Gil, M., Díaz-Peredo, E., Sánchez-Solano, F., Santos-Zavaleta, A., Martínez-Flores, I., Jiménez-Jacinto, V., Bonavides-Martínez, C., Segura-Salazar, J., Martínez-Antonio, A., and Collado-Vides, J. (2006). RegulonDB (version 5.0): Escherichia coli K-12 transcriptional regulatory network, operon organization, and growth conditions. *Nucleic Acids Research*, 34(suppl 1):D394–D397.

Santos, S. d. S., Takahashi, D. Y., Nakata, A., and Fujita, A. (2013). A comparative study of statistical methods used to identify dependencies between gene expression signals. *Briefings in Bioinformatics*, page bbt051.

Schäfer, J. and Strimmer, K. (2005). An empirical Bayes approach to inferring large-scale gene association networks. *Bioinformatics*, 21(6):754–764.

Schulz, E. G., Meisig, J., Nakamura, T., Okamoto, I., Sieber, A., Picard, C., Borensztein, M., Saitou, M., Blüthgen, N., and Heard, E. (2014). The two active X chromosomes in female ESCs block exit from the pluripotent state by modulating the ESC signaling network. *Cell Stem Cell*, 14(2):203–216.

Schwarz, B. A., Bar-Nur, O., Silva, J. C. R., and Hochedlinger, K. (2014). Nanog Is Dispensable for the Generation of Induced Pluripotent Stem Cells. *Current Biology*, 24(3):347–350.

Silva, J. and Smith, A. (2008). Capturing Pluripotency. *Cell*, 132(4):532–536.

Sing, T., Sander, O., Beerenwinkel, N., and Lengauer, T. (2005). ROCR: visualizing classifier performance in R. *Bioinformatics*, 21(20):3940–3941.

Smith, C., Roeszler, K., Hudson, Q., and Sinclair, A. (2007). Avian sex determination: what, when and where? *Cytogenetic and Genome Research*, 117(1-4):165–173.

Som, A., Harder, C., Greber, B., Siatkowski, M., Paudel, Y., Warsow, G., Cap, C., Schöler, H., and Fuellen, G. (2010). The PluriNetWork: An Electronic Representation of the Network Underlying Pluripotency in Mouse, and Its Applications. *PLoS ONE*, 5(12):e15165.

Song, L., Langfelder, P., and Horvath, S. (2012). Comparison of co-expression measures: mutual information, correlation, and model based indices. *BMC Bioinformatics*, 13(1):328.

Spivakov, M. (2014). Spurious transcription factor binding: Non-functional or genetically redundant? *BioEssays*, 36(8):798–806.

Stadler, M. B., Murr, R., Burger, L., Ivanek, R., Lienert, F., Schöler, A., Nimwegen, E. v., Wirbelauer, C., Oakeley, E. J., Gaidatzis, D., Tiwari, V. K., and Schübeler, D. (2011). DNA-binding factors shape the mouse methylome at distal regulatory regions. *Nature*.

Stolovitzky, G., Monroe, D., and Califano, A. (2007). Dialogue on Reverse-Engineering Assessment and Methods. *Annals of the New York Academy of Sciences*, 1115(1):1–22.

Sturm, O. E., Orton, R., Grindlay, J., Birtwistle, M., Vyshemirsky, V., Gilbert, D., Calder, M., Pitt, A., Kholodenko, B., and Kolch, W. (2010). The Mammalian MAPK/ERK Pathway Exhibits Properties of a Negative Feedback Amplifier. *Science Signaling*, 3(153):ra90.

Subramanian, A., Tamayo, P., Mootha, V. K., Mukherjee, S., Ebert, B. L., Gillette, M. A., Paulovich, A., Pomeroy, S. L., Golub, T. R., Lander, E. S., and Mesirov, J. P. (2005). Gene set enrichment analysis: A knowledge-based approach for interpreting genome-wide expression profiles. *Proceedings of the National Academy of Sciences of the United States of America*, 102(43):15545–15550.

Symmons, O., Uslu, V. V., Tsujimura, T., Ruf, S., Nassari, S., Schwarzer, W., Ettwiller, L., and Spitz, F. (2014). Functional and topological characteristics of mammalian regulatory domains. *Genome Research*, 24(3):390–400.

Takagi, N. and Abe, K. (1990). Detrimental effects of two active X chromosomes on early mouse. *Development*, 109(1):189–201.

Takaoka, K. and Hamada, H. (2012). Cell fate decisions and axis determination in the early mouse embryo. *Development*, 139(1):3–14.

Thévenin, A., Ein-Dor, L., Ozery-Flato, M., and Shamir, R. (2014). Functional gene groups are concentrated within chromosomes, among chromosomes and in the nuclear space of the human genome. *Nucleic Acids Research*, page gku667.

Thornhill, A. R. and Burgoyne, P. S. (1993). A paternally imprinted X chromosome retards the development of the early mouse embryo. *Development*, 118(1):171–174.

Trinklein, N. D., Aldred, S. F., Hartman, S. J., Schroeder, D. I., Otillar, R. P., and Myers, R. M. (2004). An Abundance of Bidirectional Promoters in the Human Genome. *Genome Research*, 14(1):62–66.

Trott, J., Hayashi, K., Surani, A., Babu, M. M., and Martinez-Arias, A. (2012). Dissecting ensemble networks in ES cell populations reveals micro-heterogeneity underlying pluripotency. *Molecular BioSystems*, 8(3):744–752.

Tsai, H.-K., Huang, P.-Y., Kao, C.-Y., and Wang, D. (2009). Co-Expression of Neighboring Genes in the Zebrafish (Danio rerio) Genome. *International Journal of Molecular Sciences*, 10(8):3658–3670.

Waddington, C. H. (1957). *The strategy of the genes: a discussion of some aspects of theoretical biology*. Allen & Unwin.

Wang, X., Terfve, C., Rose, J. C., and Markowetz, F. (2011). HTSanalyzeR: an R/Bioconductor package for integrated network analysis of high-throughput screens. *Bioinformatics*, 27(6):879–880.

Welling, M. and Geijsen, N. (2013). Uncovering the true identity of naïve pluripotent stem cells. *Trends in Cell Biology*, 23(9):442–448.

Wit, E. d. and Steensel, B. v. (2009). Chromatin domains in higher eukaryotes: insights from genome-wide mapping studies. *Chromosoma*, 118(1):25–36.

Witzel, F., Maddison, L. E., and Blüthgen, N. (2012). How scaffolds shape MAPK signaling: what we know and opportunities for systems approaches. *Systems Biology*, 3:475.

Woo, Y. H., Walker, M., and Churchill, G. A. (2010). Coordinated Expression Domains in Mammalian Genomes. *PLoS ONE*, 5(8):e12158.

Wu, Z. and Aryee, M. J. (2010). Subset quantile normalization using negative control features. *Journal of Computational Biology: A Journal of Computational Molecular Cell Biology*, 17(10):1385–1395.

Xiong, Y., Chen, X., Chen, Z., Wang, X., Shi, S., Wang, X., Zhang, J., and He, X. (2010). RNA sequencing shows no dosage compensation of the active X-chromosome. *Nature Genetics*, 42(12):1043–1047.

Xu, H., Ang, Y.-S., Sevilla, A., Lemischka, I. R., and Ma'ayan, A. (2014). Construction and Validation of a Regulatory Network for Pluripotency and Self-Renewal of Mouse Embryonic Stem Cells. *PLoS Comput Biol*, 10(8):e1003777.

Xu, H., Baroukh, C., Dannenfelser, R., Chen, E. Y., Tan, C. M., Kou, Y., Kim, Y. E., Lemischka, I. R., and Ma'ayan, A. (2013). ESCAPE: database for integrating high-content published data collected from human and mouse embryonic stem cells. *Database*, 2013(0):bat045–bat045.

Xue, K., Ng, J.-H., and Ng, H.-H. (2011). Mapping the networks for pluripotency. *Philosophical Transactions of the Royal Society B: Biological Sciences*, 366(1575):2238–2246.

Yaffe, E., Farkash-Amar, S., Polten, A., Yakhini, Z., Tanay, A., and Simon, I. (2010). Comparative Analysis of DNA Replication Timing Reveals Conserved Large-Scale Chromosomal Architecture. *PLoS Genet*, 6(7):e1001011.

Yi, F., Pereira, L., and Merrill, B. J. (2008). Tcf3 Functions as a Steady-State Limiter of Transcriptional Programs of Mouse Embryonic Stem Cell Self-Renewal. *STEM CELLS*, 26(8):1951–1960.

Yildirim, E., Sadreyev, R. I., Pinter, S. F., and Lee, J. T. (2012). X-chromosome hyperactivation in mammals via nonlinear relationships between chromatin states and transcription. *Nature Structural & Molecular Biology*, 19(1):56–61.

Ying, Q.-L., Wray, J., Nichols, J., Batlle-Morera, L., Doble, B., Woodgett, J., Cohen, P., and Smith, A. (2008). The ground state of embryonic stem cell self-renewal. *Nature*, 453(7194):519–523.

Yuan, Y., Li, C.-T., and Windram, O. (2011). Directed Partial Correlation: Inferring Large-Scale Gene Regulatory Network through Induced Topology Disruptions. *PLoS ONE*, 6(4):e16835.

Zalesky, A., Fornito, A., and Bullmore, E. (2012). On the use of correlation as a measure of network connectivity. *NeuroImage*, 60(4):2096–2106.

Zhang, H.-M., Chen, H., Liu, W., Liu, H., Gong, J., Wang, H., and Guo, A.-Y. (2011). AnimalTFDB: a comprehensive animal transcription factor database. *Nucleic Acids Research*.

Zhang, X., Zhao, X.-M., He, K., Lu, L., Cao, Y., Liu, J., Hao, J.-K., Liu, Z.-P., and Chen, L. (2012). Inferring gene regulatory networks from gene expression data by path consistency algorithm based on conditional mutual information. *Bioinformatics*, 28(1):98–104.

Zhu, H., Rao, R. S. P., Zeng, T., and Chen, L. (2012). Reconstructing dynamic gene regulatory networks from sample-based transcriptional data. *Nucleic Acids Research*, 40(21):10657–10667.

Zoppoli, P., Morganella, S., and Ceccarelli, M. (2010). TimeDelay-ARACNE: Reverse engineering of gene networks from time-course data by an information theoretic approach. *BMC Bioinformatics*, 11(1):154.

Zvetkova, I., Apedaile, A., Ramsahoye, B., Mermoud, J. E., Crompton, L. A., John, R., Feil, R., and Brockdorff, N. (2005). Global hypomethylation of the genome in XX embryonic stem cells. *Nature Genetics*, 37(11):1274–1279.

6 Materials and Methods

6.1 Array Analysis for the three cell lines XO,XY and XX

R/bioconductor tools Microarrays were background corrected using RMA (Robust Multi-array Analysis) as available in the bioconductor affy package (http://www.bioconductor.org) and normalized using the custom subset quantile normalization (see below). Probeset annotation was provided by the xmapcore database for Ensembl mouse version 62 (http://annmap.picr.man.ac.uk/download/).

Custom subset quantile normalization We use subset quantile normalization (SQN) (Wu and Aryee, 2010) in order to exclude X-linked genes from impacting the quantile normalization. This is done because we compare cell lines with different sex chromosome complement. The implementation of SQN by Wu et al. expects a mixing parameter w that we chose as $w = 0$. Values $w > 0$ are chosen to avoid truncation of the reference distribution if it is calculated from a small subset of probes. Since we use all autosomal probesets and these probesets form the majority of probesets, truncation issues are negligible. To define the set of probes used by SQN, we excluded probesets mapping to genes from chromosomes X and Y as well as mitochondrial DNA and probesets that cannot be mapped any chromosome. Thus we use only probesets mapped to autosomal genes and control probesets.

Annotation of genes For each gene the expression is calculated by averaging over probesets that detect the maximum number of transcripts per gene. Pseudogenes as well as probesets annotated in xmapcore as unreliable or non-exonic were removed.

6.2 Chapter 2

Defining ancestral and newly acquired genes for the X chromosome and autosomes We followed the strategy of He et al. (2011) to identify X-linked and

Table 6.1: Number of protein coding genes detected by the Exon array and the RNA-seq data falling into the categories of autosomal/X-linked genes and ancestral/newly acquired genes. The column titles % indicate the fraction between detected genes and annotated genes of the indicated category.

Autosomal	Ancestral	Ensembl	Array	% Array	RNA-seq	% RNA-seq
FALSE	FALSE	679	413	0.61	221	0.33
FALSE	TRUE	261	237	0.91	176	0.67
TRUE	FALSE	12957	9416	0.73	5600	0.43
TRUE	TRUE	7939	7539	0.95	5958	0.75

autosomal genes that were already present on the chromosomes of the common ancestor of mouse and chicken (Julien et al., 2012). These genes are called ancestral in the main text. Chicken homology information for mouse genes were downloaded from the Ensembl database v79 (www.ensembl.org) using the R package biomaRt. To determine orthologs, we restricted to protein coding genes that were qualified by

```
ggallus_homolog_orthology_confidence=1 and
ggallus_homolog_orthology_type=ortholog_one2one
```

in the database. X-linked mouse genes with chicken orthologs were then determined by filtering all orthologs for the ones that are found on the chicken chromosomes 1 and 4. For autosomal mouse genes, all orthologs were taken except the ones on chicken chromosomes 1 and 4. The number of ancestral and acquired genes detected by the platforms used to assay the transcriptome are shown in Tab. 6.1.

RNA-seq data supporting upregulation of ancestral X-linked genes   Normalized expression values of RNA-seq data for male and female mouse embryonic stem cells and male and female neuronal progenitor cells were obtained from the supplementary information of Gendrel et al. (2014). The table S1 available at http://www.sciencedirect.com/science/article/pii/S1534580714000574 was downloaded and gene symbols were converted to Ensembl gene ids using the R package biomaRt with the mouse genome at Ensembl version 79, available at www.ensembl.org. Only genes with biotype protein_coding in the Ensembl database were retained. The FPKM values provided by the authors were transformed using the function $\log2(x+1)$.

Identification of genes regulated over time   The null model was generated by shuffling the time order of the expression values obtained for a given gene 10000 times, each time smoothing the time series (using `SMA_lin`, which estimates a running average for time series with non-equidistant time points, with time window of 24 h (Eckner, 2012)) and calculating the variance of the smoothed time series.

We scored the genes by counting how often the variance of the original smoothed time series exceeds the variance of the shuffled time series. We termed those genes as regulated that obtained the highest score for either the XO or XX time series, i.e. which showed more variance than any permuted time series in the respective time series (corresponding to $p < 10^{-4}$).

Definition of delay score and GSEA delay analysis   To compute a score measuring delay, we further restrict the set of regulated genes to those genes that have a difference smaller than 1 in log2 expression in undifferentiated XO and XX mESC. The delay score compares rates of change of the kinetics of a given gene in the XO and XX time series, thus a positive delay score can be interpreted as a larger rate of change in XO than in XX. The XX and XO time series were smoothed for each gene with `SMA_lin` using a time window of 24 h. These time series are then linearly interpolated. Finally, the delay score is computed by taking the average difference of the absolute numerical time derivative of the smoothed interpolated XO and XX time series. This corresponds to the formula

$$DS(g(t)) = \left| \frac{d}{dt} g_{XO}(t) \right| - \left| \frac{d}{dt} g_{XX}(t) \right|, \qquad (6.1)$$

where $g_{XO}(t)$ and $g_{XY}(t)$ denote the smoothed kinetics of gene g in the XO and XY time series.

To test for enrichment of gene sets with common Gene Ontology biological process (GOBP, including their descendants with relation `is_a` and `part_of`) annotation, we used the gene set enrichment analysis method (GSEA) (Subramanian et al., 2005) as implemented in the HTSanalyzeR bioconductor package (Wang et al., 2011). The GOBP annotations are taken from the org.Mm.eg.db and GO.bp bioconductor packages (dababase time stamp 03/2012). We restricted the tested gene sets to a minimum size of 5 genes and used 12000 permutations.

We find only two terms significantly enriched at an FDR $< 5\%$, "stem cell development" and "stem cell maintenance". For a list of the top enriched terms, see Tab. 6.2.

Table 6.2: Top enriched GO terms in GSEA analysis

Rank	GO_ID	Term	FDR
1	GO:0048864	stem cell development	0.010
2	GO:0019827	stem cell maintenance	0.019
3	GO:0048863	stem cell differentiation	0.138
4	GO:0001708	cell fate specification	0.170
5	GO:0051146	striated muscle cell differentiation	0.389
6	GO:0002377	immunoglobulin production	0.420
7	GO:0051272	pos. reg. of cellular component movement	0.421
8	GO:0040017	pos. reg. of locomotion	0.422
9	GO:0065004	protein-DNA complex assembly	0.425
10	GO:0032387	neg. reg. of intracellular transport	0.428

Identification of differentially expressed genes

First, we defined the set of genes that are expressed highly enough to be reliably detected in at least one of the three cell lines, using a threshold, which we determined by comparing the expression of X-linked genes between the XO and XX cell lines before differentiation. As shown in Fig. 6.1, page 133, X-linked genes are expressed on average at twofold higher levels in XX than in XO ES cells at time 0 h for expression values higher than 7.5, which we identified as threshold below which the microarray responds sub-linearly to concentration changes. Differentially expressed genes were defined by comparing the difference of gene expression between the XX and XO or XY cell lines to an appropriate null model, which was calculated using biological replicates of the expression of undifferentiated ES cells.

More precisely, we computed an empirical error model $s(x)$ that relates the standard deviation s to the expression x of a gene. This was constructed using mean expression and standard deviation calculated for each gene using the three duplicates at time point 0 h. From this, an average standard deviation in each expression bin of width 0.5 was calculated, and the resulting values were interpolated using loess with a span of 3. Then, the difference in expression was divided by the value of the error model for their average expression, yielding z-scores. To estimate an upper limit for the false discovery rate (FDR) we calculated a z-value null distribution by applying the same procedure for expression differences for pairs of replicates for the three cell lines XX, XO and XY.

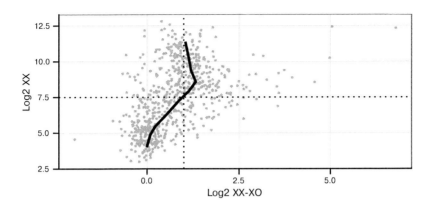

Figure 6.1: Microarrays resolve dosage differences linearly only above the expression threshold 7.5 (horizontal broken line). The expression of all X-linked genes in undifferentiated XX cells was plotted against the difference in expression between undifferentiated XX and undifferentiated XO cells. The vertical broken line indicates a logarithmic fold change of 1, corresponding to a linear fold change of 2. The black line was constructed by computing the expression mean in the XX cell line and the mean difference XX-XO in ten equally sized expression bins according to expression in the XX cell line.

To test for differential expression, we compared our z-values with the z-value null distribution with the largest standard deviation to ensure that we estimate an upper bound for the FDR in the following. Genes were called significantly differentially expressed if their z-value exceeded 5. For X-linked genes, we expect a 2-fold difference in expression, and thus we tested for differential regulation by shifting the expression in XX by one logarithmic unit using the same critical z-value zcrit=5. Using these criteria for X-linked and autosomal genes, we found a total number of 1214 differentially expressed genes at an FDR estimated below 11.6 %.

GSEA of de novo methylated genes To test whether delayed genes gain *de novo* methylation during differentiation, GSEA was performed using a set of *de novo* methylated genes, based on genome-wide methylation profiles of ES cells and differentiated neural precursor cells obtained by Stadler et al. (2011). Regions that were unmethylated (UMR) in ESCs and fully methylated (FMR) in neural precursor cells were defined as *de novo* methylated. If a 100 bp region (-50 bp to 50 bp) surrounding a gene's transcription start site overlaps with a *de novo* methylated region, the gene was defined as *de novo* methylated.

Enrichment of differentially expressed genes among DNA methylation sensitive genes We used three publicly available data sets where the expression profile of Dnmt1-/-Dnmt3a-/- Dnmt3b-/- triple knock out XY ES cells (TKO) was compared to the parental wildtype J1 line:

- The list of genes (overlap with our set of expressed genes: N=75) upregulated in TKO cells identified by Fouse et al. (2008) (Table S5, column "1st Round Analysis Upregulated in TKO Gene List")

- Genes found upregulated in a TKO RNA-seq transcriptome analysis from Karimi et al. (2011) (Table S1, sheet 3 "DNMT TKO UP", overlap with our set of expressed genes: N=119)

- TKO transcriptome analysis using microarrays from Sakaue et al. (2010), for which the raw data was downloaded from the GEO data base (with accession number GSE20177) and normalized using rma from the affy bioconductor package. Only probesets that had mas5 present calls in at least 2 wild type and TKO replicates were used to exclude weakly expressed genes. A z-value null distribution as described above was calculated, and genes were termed upregulated if the difference between TKO ES and wild type ES expression lead to a z-value larger than 6, yielding an estimated FDR of 10.20%. Overlap with our set of expressed genes: N=566.

Distance heat map and PCA for genes regulated over time For the distance heat map, we first smoothed all gene kinetics with SMA_lin using a time window of 24 h and linearly interpolated the resulting time series for each gene to arrive at an expression value for every hour between 0 h and 84 h. From this data we calculated the heat map, which shows the normalized (linearly scaled to range between 0 and 1) Euclidean distance between the expression vectors corresponding to the indicated cell lines and time points. The red line connects the points on lines perpendicular to the diagonal with the minimal distance between the two time series and is interpolated using the R function loess with a span of 0.6.

For the PCA, we generated a list of genes annotated as stemcell maintaining (GO:0019827) using the GO.db bioconductor package. We intersected this list with the genes that we termed as regulated. Expression in all three time series XX, XO and XY was analyzed with the principal component analysis (PCA), taking the

Table 6.3: List of genes annotated as stemcell maintaining factors found to be regulated in our transcriptome time series. The contribution to the first two principal components of the PCA is shown for the subset of genes that was included in the PCA analysis.

Gene	PC1	PC2
Hes1	-0.06	-0.35
Kit	-0.24	-0.38
Klf4	-0.35	-0.06
Ascl2	0.07	0.19
Sox2	-0.16	0.16
Sox4	0.26	0.03
Tbx3	-0.44	-0.28
Tcfap2c	-0.16	-0.29
Tcl1	-0.44	0.41
Ctr9	0.03	0.15
Yap1	0.04	0.05
Esrrb	-0.46	0.41
Med14	0.08	0.20
Nanog	-0.21	0.05
Lin28a	0.20	0.32

different time points as observations in an n-dimensional expression state space, where n=15 is the number of stemcell maintaining factors analyzed. For a list of genes included in the analysis, see Tab. 6.3.

6.3 Chapter 3

Gene position annotation for X inactivation center computation Gene position annotation was obtained using the xmapcore database for Ensembl mouse version 62 (`http://annmap.picr.man.ac.uk/download/`) based on the NCBI m37 genome assembly. Genes were filtered to exclude pseudogenes from the analysis, but non-coding genes were retained.

TAD annotation for 5C data The TAD borders were determined by hand using the normalized contact map (Fig. 3.1, page 55). The list of TAD border coordinates is given in Tab. 6.4.

Table 6.4: Annotation of the TADs around the X inactivation center

Chromosome	Start	End	TAD label
chrX	98831148	99362574	TAD#A
chrX	99362574	99422675	TAD#B
chrX	99440000	99840000	TAD#C
chrX	100297658	100663727	TAD#D
chrX	100663727	101182969	TAD#E
chrX	101271697	102192371	TAD#F
chrX	102192371	102325818	TAD#G
chrX	102325818	102998976	TAD#H
chrX	102998976	103425147	TAD#I

Hi-C data   Normalized Hi-C data published in (Dixon et al., 2012) was downloaded from http://promoter.bx.psu.edu/hi-c/download.html (Mouse ES Cell Raw Matrices).

Gene pairs were sorted into six bins according to their contact frequency relative to similarly distant gene pairs. Gene pairs were binned into equidistant distance bins of 500 kb width. Subsequently gene pairs in the same distance bin were binned into 6 equally sized bins according to contact frequency, with 1 corresponding to the lowest bin and 6 to the highest.

TAD annotation for Hi-C data   The Ren TAD annotation for the mm9 genome assembly was downloaded from http://promoter.bx.psu.edu/hi-c/download.html (Mouse ES cell topological domains).

The Kingsford TAD annotation was obtained using the Armatus software which is based on an algorithm presented in (Filippova et al., 2014). The Armatus software was downloaded from https://github.com/kingsfordgroup/armatus on January 22, 2015. Armatus was applied to the normalized Hi-C data with parameters g=0.5 (maximal resolution γ_{max}), s=0.005 (step size) and k=50 (number of top optimal solutions).

Gene position annotation and computation of correlation for genome-wide computation   Gene positions for the mouse genome were obtained from the UCSC genome browser (http://genome.ucsc.edu/index.html) using the GenomicFeatures R package. The mm9 genome assembly was used together with the Ensembl version supplied by the UCSC genome browser (Ensembl data time stamp 14.11.2011).

Genes from chromosomes 1-19 and X were filtered to retain only protein-coding genes. Correlation between pairs of genes from the same chromosome was computed by taking the average of the correlation in the XO, XY and XX time series.

Expression data for randomly integrated reporters Raw count data and mm9 reporter locations for mPGK and tet-Off constructs were downloaded from the supplementary information of Akhtar et al. (2013) at http://www.sciencedirect. com/science/article/pii/S0092867413008891. Expression counts were divided by normalization counts to arrive at normalized expression values and the mean of replicates was calculated. For the tet-Off constructs, expression data is available at multiple inductions. The expression for each reporter in this case was calculated by taking the log2 of the normalized expression values corresponding to an induction with 100ng and 0.1ng of Doxycyclin. Subsequently, the average of the logarithmized expression of both inductions was calculated. Reporters with zero counts at one of the inductions were omitted. The expression data for the mPGK construct was similarly logarithmized.

PPI data from StringDB Data from StringDB for mouse at version 9.1 was obtained using the R package STRINGdb. Scores from three types of evidence was obtained: textmining, database, and experimental. From these three scores, an average score was computed for each pair of proteins and the result was logarithmized (log2). Ensembl protein ids were converted to Ensembl gene ids using the Ensembl database (www.ensembl.org) at version 78.

6.4 Chapter 4

Algorithms and packages used An overview of the algorithms used and the publications describing these algorithms is given in Tab. 6.5. The two partial correlation algorithms used in this work are based on partial least squares (pls) and a lasso regression (lasso). The functions supplied by the parcor package for both algorithms use cross-validation for determining the optimal partial correlation prediction model (Krämer et al., 2009). The parameter k determines the number of splits for this cross-validation and should be set as large as possible. To limit the computation time this parameter had to be set to $k = 3$. It is not listed in the table because it is not explicitly linked to the reconstruction strategy. For the ARACNE algorithm, we used the aracne.m command in parmigene.

Table 6.5: Overview of algorithms and packages employed. Only the ARACNE algorithm requires a parameter that indicates the stringency of the link pruning (Sales and Romualdi, 2011).

Score	Publication	Package	Parameters
Pearson correlation		R base	
Spearman correlation		R base	
Mutual information		parmigene	
ARACNE	Margolin et al. (2006)	parmigene	τ
CLR	Faith et al. (2007)	parmigene	
MRNET	Meyer et al. (2007)	parmigene	
Partial correlation	Krämer et al. (2009)	parcor	

Gold Standards A list of transcription factors in mouse was obtained from the Animal Transcription Factor Database (ATFDB) (Zhang et al., 2011), located at http://www.bioguo.org/AnimalTFDB/. We downloaded the list named 'Gene list of TFs', providing transcription factor Ensembl gene ids. For each of the following gold standards, TFs were determined using this list.

ChIP-seq benchmark

ChiP-seq data were downloaded from the Mouse ES Cell ChIP-Seq Compendium maintained by the Bioinformatics Core at Wellcome Trust - MRC Stem Cell Institute, University of Cambridge, http://lila.results.cscr.cam.ac.uk/ES_Cell_ChIP-seq_compendium_UPDATED.html. Subsequently, for each ChIP-seq experiment, we computed the transcription factor association score (TFAS) for each gene. The TFAS for gene i is defined in (Ouyang et al., 2009) as $TFAS_i = \sum_k g_k \exp(-d_{ki}/d_0)$, where k runs over all peaks determined by ChIP-seq, g_k is the corresponding peak height, d_{ki} is the distance between the peak mid-point and the transcription start site of gene i and d_0 is a characteristic length scale which we set equal to 1000 bp. A list of source data files for each ChIP-seq data set is given in Tab. 6.6, page 139.

Overexpression benchmark

Expression data was obtained from the GEO database by downloading the series GSE31381 and GSE14559 using the R package GEOquery. Because these series contain a mix of already logarithmized expression values and linear expression values, all samples with median greater than 10 were logarithmized. The validity of this workaround was tested by comparing the standard deviation of all sam-

Table 6.6: Overview of the data sources for the ChIP-seq data sets used for individual transcription factors.

TF	Filename
Aff4	`AFF4_Shilatifard_GSE30267.bw`
Ctcf	`CTCF_Buchholz_GSE24030.bw`
E2f1	`E2f1_Ng_GSE11431.bw`
Setdb1	`Eset_Ng_GSE17642.bw`
Esrrb	`Esrrb_Ng_GSE11431.bw`
Klf4	`Klf4_Ng_GSE11431.bw`
Nanog	`Nanog_Ng_GSE11431.bw`
Pou5f1	`Oct4_Lemischka_GSE22934.bw`
Prdm14	`Prdm14_Wysocka_GSE25409.bw`
Rest	`REST_Young_GSE26680.bw`
Stat3	`STAT3_BrgKO_Crabtree_GSE27708.bw`
Sox2	`Sox2_Ng_GSE11431.bw`
Tcf3	`Tcf3_Young_GSE11724.bw`
Tfcp2l1	`Tcfcp2l1_Ng_GSE11431.bw`
Zfx	`Zfx_Ng_GSE11431.bw`
Mycn	`n-Myc_Ng_GSE11431.bw`

ples after logarithmizing. The standard deviation was observed to be very similar sd=1.225 ± 0.019.

To obtain a z-value null distribution, we first construct an empirical error function $\sigma(x)$ that gives the mean standard deviation as a function of expression x. We calculated the mean expression and the standard deviation for each gene using the four samples labeled as 'Control'. The empirical error function was defined by a Loess fit on the standard deviation vs. the mean with a span of 3.

We then computed the z-value for each gene and each transcription factor. For each transcription factor, the average of the gene expression for the non-induced condition (Dox+) was substracted from the average for the induced condition (Dox−). The result was divided by the empirical error function evaluated at the mean of the induced and non-induced conditions.

We identified differentially expressed genes at a false discovery rate < 0.001 by imposing an absolute z-score threshold of 2.9. Additionally we required a fold change of more than 2.

LoF benchmark

The loss of function benchmark data was obtained from the Escape database (Xu et al., 2013). The list of differential genes for each transcription factor assayed was

downloaded from `http://www.maayanlab.net/ESCAPE/download/logof.txt.zip`.
This data set contains different loss of and gain of function experiments with transcription factors in mESC, using thresholds of different stringency for identifying
differential genes. We kept only experiments marked as loss of function. Gene
symbols were translated to Ensembl gene ids using the symbolToGene function of
the R package annmap (Ensembl version 74).

Kd benchmark

The transcription factor knockdown data was obtained from the publication
Nishiyama et al. (2013). We downloaded the list of differential genes for each
transcription factor assayed from `http://www.nature.com/srep/2013/130306/`
`srep01390/extref/srep01390-s1.xls`. Gene symbols were translated to Ensembl gene ids using the symbolToGene function of the R package annmap (Ensembl version 74).

Literature-based PluriNetWork   The literature based PluriNetWork described in
Som et al. (2010) was downloaded from WikiPathways (`http://wikipathways.`
`org/index.php/Pathway:WP1763`) on April 15, 2014 using Cytoscape 3.1.0 and
exported as plain text file. It was subsequently imported into R and restricted to
node pairs that could be mapped to Ensembl gene ids using Ensembl version 79.
We furthermore retained only interactions where the first node was a transcription
factor, yielding 362 interactions.

Transcriptome data   Microarray samples were obtained from the GEO database
automatically using the R packages GEOquery for data retrieval and GEOmetadb
for data selection. Using GEOmetadb (time stamp 11/2013), the GEO database
was searched for entries corresponding to the Affymetrix Mouse Gene St platform
(GPL6246). These entries were filtered by the following criteria. First, the raw
CEL data had to be present. Second, the description had to contain at least one of
the following keywords: mESC, stem cell, stem cells, Oct4, Sox2, Nanog, Pou5f1,
embryonic. The corresponding 1194 samples were then downloaded. A list of GEO
series ids corresponding to the obtained data sets is given at the end of this chapter.

These samples were normalized together using the rma function from the R package oligo. Probesets were annotated with Ensembl gene ids using the R package
mogene10sttranscriptcluster.db. Probesets that were associated with more than
one gene were omitted. The final expression matrix contained 19615 genes measured in 1194 samples.

To compute the correlation of gene expression with that of randomly integrated
reporters (IR), we proceeded as follows. The expression of IRs was computed as

in Sec. 6.3. For each gene, the average expression of all IRs closer than 10 kb
was determined. If no IRs could be found below this distance, the respective
gene was omitted from the correlation analysis. Next, for each microarray sample,
Spearman's correlation was computed between the gene expression in this sample
and the matching average IR expression of the closest IRs.

Gene positions for the mouse genome were obtained from the UCSC genome
browser (http://genome.ucsc.edu/index.html) using the GenomicFeatures R
package. The mm9 genome assembly was used together with the Ensembl version
supplied by the UCSC genome browser (Ensembl data time stamp 14.11.2011).

The list of GEO series obtained through keyword filtering was:

GSE14288, GSE14411, GSE14412, GSE16428, GSE16429, GSE16430, GSE17004,
GSE18009, GSE18074, GSE18393, GSE18691, GSE19377, GSE19378, GSE19542,
GSE19604, GSE20220, GSE21062, GSE21595, GSE22946, GSE23033, GSE23341,
GSE23406, GSE23547, GSE23956, GSE23957, GSE24046, GSE24929, GSE25255,
GSE26624, GSE26653, GSE26850, GSE27079, GSE27087, GSE27485, GSE27685,
GSE28710, GSE29094, GSE29347, GSE29590, GSE30076, GSE30244, GSE30428,
GSE30444, GSE30445, GSE30537, GSE30541, GSE31456, GSE31523, GSE31562,
GSE31784, GSE32098, GSE32994, GSE33091, GSE33502, GSE33953, GSE34060,
GSE34232, GSE34795, GSE35063, GSE35110, GSE35111, GSE35395, GSE35775,
GSE36096, GSE36897, GSE37113, GSE37548, GSE37832, GSE39103, GSE39321,
GSE39637, GSE39955, GSE40323, GSE40324, GSE40327, GSE40335, GSE40582,
GSE40701, GSE41133, GSE41171, GSE41298, GSE41597, GSE42210, GSE42250,
GSE42782, GSE43133, GSE43387, GSE43398, GSE43420, GSE43421, GSE43613,
GSE43850, GSE43961, GSE44107, GSE44972, GSE45352, GSE45826, GSE45837,
GSE45916, GSE46250, GSE46319, GSE46405, GSE46860, GSE47067, GSE47161,
GSE48092, GSE48146, GSE48315, GSE48411, GSE48438, GSE49787, GSE50017,
GSE51194, GSE51960

.

Calculation of AUPR   For each interaction score, TF-gene pairs were ordered by
the absolute value of the interaction score. For each gold standard, each TF-gene
pair was then labeled as true or false according to whether it was present in the
gold standard or not. Using the ranked prediction list and the labels, the precision-
recall curve can be computed. This was done using the interface of the ROCR
package (Sing et al., 2005). Since it is not implemented in ROCR, custom code
for computing the area under the precision recall curve (AUPR) was plugged into
the ROCR interface. The custom code uses the interpolation algorithm of Boyd

Table 6.7: Quantification of topological properties of the predicted networks using igraph.

Quantity	igraph function
% in largest component	clusters
no connected components	no.clusters
modularity edge betweenness	modularity, edge.betweenness.community
modularity fastgreedy	modularity, fastgreedy.community
diameter	diameter
transitivity	transitivity
degree correlation (assortativity)	assortativity
degree distribution	degree.distribution

et al. (2013), available at `https://github.com/kboyd/raucpr`. This interpolation algorithm was suggested by Davis and Goadrich (2006).

The obtained AUPRs were used to calculate the improvement of the predictions over random. The improvement over random was obtained by dividing the AUPR for each interaction score by the theoretical precision of a random score. This is given by the number of pairs labeled as true divided by the total number of pairs. If the AUPR was only computed up to a recall of 0.05, the resulting random precision also has to be multiplied by 0.05 to arrive at the random AUPR.

Quantifying topological properties of the literature network and the predicted networks   All topological properties were computed using the R package igraph. For an overview of the functions used, see Tab. 6.7.

7 Supplementary Figures

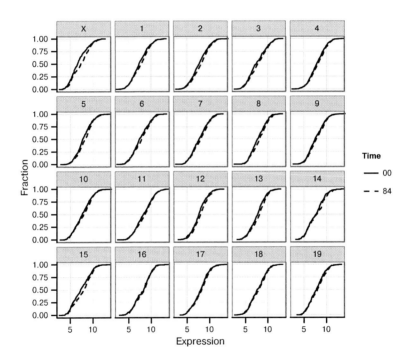

Figure 7.1: Cumulated density functions for ancestral genes from all autosomal chromosomes and the X chromosome, comparing expression at 0 h with expression at 84 h in the XO cell line.

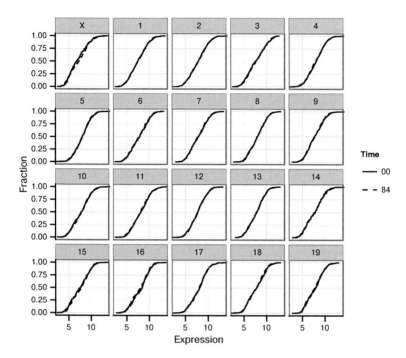

Figure 7.2: Cumulated density functions for ancestral genes from all autosomal chromosomes and the X chromosome, comparing expression at 0 h with expression at 84 h in the XY cell line.

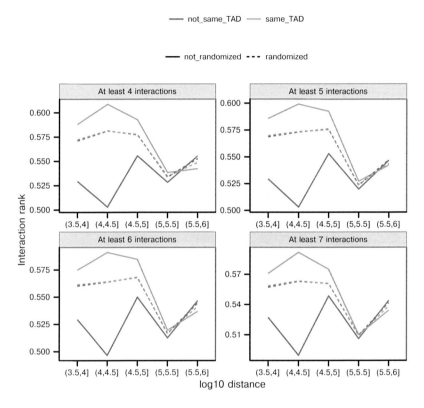

Figure 7.3: The preference of protein-coding genes for binding partners in the same TAD is conserved when restricting String DB evidence to small-scale experiments and annotations. Gene pairs are binned with respect to equidistant logarithmic distance intervals. For each gene, the interaction rank of its binding partners is computed from String DB scores labeled as database (validated small-scale interactions, protein complexes, and annotated pathways) and average scores are computed separately for interaction partners in the same TAD (blue line) and interaction partners not in the same TAD (red line). The same computation is repeated for 250 reshufflings of random domains with same size and gap size distribution as the real TADs (broken lines). Different panels contain data with the minimum number of interactions per gene indicated, ranging from a minimum of 4 to a minimum of 7 interaction partners.

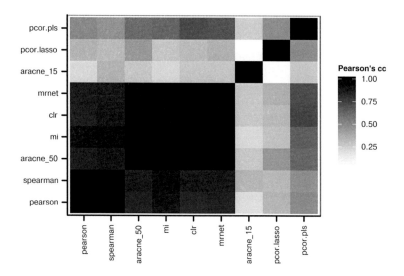

Figure 7.4: Correlation of the improvement of the $AUPR_{0.05}$ over random for individual TF predictions between the different scores. The heat map shows the mean of the correlation values for the four gold standards ChIP-seq, Kd, LoF and Overexpression.

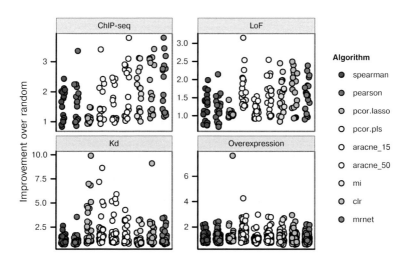

Figure 7.5: Improvement of the $\mathrm{AUPR}_{0.05}$ over random predictions for individual transcription factors. Improvements are shown for four gold standards based on ChIP-seq measurements (ChIP-seq), differential gene expression after TF knockdown (Kd), TF loss of function experiments (LoF) and differential gene expression upon TF overexpression (Overexpression). Algorithms used for the comparison are Spearman correlation (spearman), Pearson correlation (pearson), mutual information (mi), ARACNE (with cutoff parameter 0.15 and 0.5), CLR, MRNET and partial correlation in the pls (pcor.pls), lasso (pcor.lasso) implementation. Note that for the Kd gold standard, three data points with improvement > 10 have been omitted to improve the visibility of differences.